The latest scientific information about
the dangers of not treating "Low T"

# If your testosterone is low,

# you're gonna get fat.

**+17** other things the FDA doesn't want
you to know about testosterone.

## DR. CHARLES MOK

ALLURE MEDICAL PUBLISHING
SHELBY TOWNSHIP, MI

Published by
Allure Medical Publishing
Shelby Township, MI

      Publisher's Cataloging-in-Publication Data
      Mok, Charles.

      If your testosterone is low, you're gonna get fat : +17 other things
      the FDA does not want you to know about testosterone / Charles
      Mok. – Shelby Township, MI : Allure Medical Pub., 2018.

      p. ; cm.

      ISBN13: 978-0-9980195-0-5

      1. Testosterone—Therapeutic use—United States.
      2. Testosterone—Health aspect. I. Title.

      QP572.T4 M85 2018
      612.61—dc23                          2017956052

FIRST EDITION

Project coordination by Jenkins Group, Inc.
www.BookPublishing.com

*Interior design by Brooke Camfield*

Printed in the United States of America
22   21   20   19   18  •  5   4   3   2   1

# Contents

# Acknowledgments

I would like to thank individuals who helped me write this book. It is by no means a complete list, but there are notable mentions.

Michelle Byars has been working in our natural hormone replacement team for almost 10 years. She is a medical assistant and assists the doctors, physician assistants, and nurse practitioners, but she really runs the entire department. She asked me to help make tools and learning modules so she could get new providers up to speed on what we offer and why we do it.

Mary Scott, Angela Heitman, and Tiffany Wisnieski have been my team for getting this and other publications completed and readable. They have helped make this book an attractive, interesting, and easy-to-digest review of the medical literature.

Mary Carr, Master of Library Science at McLaren Macomb Medical Center, assisted me in obtaining the numerous clinical research papers that I needed to explain these cutting-edge concepts.

And my wife of 31 years, Natalie, understood why I needed print, and not electronic, versions of clinical research cluttering the den and other areas of our home. She also understood that "deep work time" meant I wasn't going to be tuned in to anything else going on at the time.

# Introduction

<u>**Men, If Your Testosterone Is Low, You're Gonna Get Fat!**</u>
I'm looking forward to telling you the story of the science behind testosterone replacement. You may have heard the low-testosterone marketing
campaigns ("Low T") . . . and then noticed they disappeared. There's a
specific reason for that—in 2014, the US government changed its rules
regarding testosterone replacement therapy (TRT or TR).

Testosterone replacement therapy has been around for decades and
was originally intended to treat men who, for various reasons, did not
develop masculine characteristics. For example, there are certain genetic
disorders where a man has an extra female chromosome, so as he grows
up, he may develop effeminate features and lack masculine features, be
on the small side, and have very little body hair as a result of failure to
produce adequate testosterone. These men would be given testosterone.
Their health would improve, and they would develop some masculine
characteristics.

Testosterone is produced in the testicles in response to hormonal
stimulation from the brain. The gland in the brain that provides this signal
is called the pituitary gland. Men cease producing testosterone when their
pituitary gland or testicles are injured, damaged, or diseased. When men
have low testosterone, they will experience various health problems, such
as obesity, diabetes, heart disease, lack of sex drive, and/or depression.

When testosterone replacement is used to treat these men, however, their symptoms generally resolve.

The historical reason for using testosterone replacement has been to treat this particular group of men (i.e., with organ damage or genetic disorders), but as time went on, we realized that some men develop low testosterone even when there isn't specific organ damage or a genetic disorder. In fact, beginning at age 30, testosterone starts to decline every year at a rate of 1%–3%. Some men's levels decline faster; some decline slower. When doctors were presented with patients who had symptoms associated with low testosterone but no organ damage, they prescribed testosterone replacement therapy and noticed that their patients got better.

Starting around the year 2000, the rate of testosterone prescriptions went up dramatically as more and more men have realized that the things they associated with getting older could be improved with testosterone replacement. With more public awareness of these benefits and more men seeking treatment, physicians became more attuned to low-testosterone symptoms and started routinely checking men's testosterone levels. They also started prescribing testosterone replacement therapy when appropriate. In an effort to raise public awareness of TRT, drug companies started promoting testosterone directly to consumers. The "Low T" marketing campaigns were everywhere—TV, radio, Internet, and magazines.

For reasons we will talk about in the first chapter, the FDA put a stop to that, changing the rules for testosterone prescriptions to indicate that they could be used only in a unique, narrow set of circumstances and stating that testosterone should not be used for men with low-testosterone syndrome if its cause could not be identified. Specifically, the FDA said that although the agency had noted that men with symptoms of low testosterone were being successfully treated with TRT, they should not have been treated in the first place because their low testosterone was secondary to aging and not caused by specific organ damage or a genetic disorder.

As absurd as this scenario sounds, it's true: the FDA singled out low testosterone as being a disease that should not be treated because it occurs more frequently as men age. Physicians are allowed (and encouraged) to treat diabetes, cardiovascular disease, cancer, arthritis, high blood pressure, and other diseases that are definitively associated with aging, but the FDA ruled that since testosterone gets low in men as they age, this condition should not be treated—men should just accept low testosterone as a consequence of aging and live with it.

The title of this book, *If Your Testosterone is Low, You're Gonna Get Fat*, is meant to catch your eye. And it's a true statement—men with low testosterone typically gain weight, while men with normal testosterone levels typically maintain or lose weight. Men with low testosterone who are obese will almost always lose weight when put on testosterone. Not only that, they'll keep the weight off for years.

But aside from weight gain, low testosterone is associated with a myriad of diseases: it's specifically and definitively linked to diabetes, obesity, metabolic syndrome, cardiovascular disease (including heart attacks and strokes), depression and other psychological problems, decreased sexuality and libido, and erectile dysfunction. Most important, in large-scale, long-term studies, untreated low testosterone is associated with a doubled chance of premature death.

I am excited to write and share this book. The timing is perfect! I recently wrote a book on a cutting-edge treatment for women's health via the application of hormones (*Testosterone: Strong Enough for a Man, Made for a Woman*), and although I treat many women with hormonal imbalances, I've been treating almost as many men who likewise have a hormone imbalance. Let me tell you more about why the timing is great for me to write this book and tell this interesting story.

I run a medical practice with about 200 team members and about 20 doctors, physician assistants, and nurse practitioners. We opened in 2004 and grew very fast to where we are now. In the early days, I focused my time on innovation, finding niches that were not being served, and constantly learning and leading. As time went on, however, patient care swallowed up my time. In March of 2016, I decided to phase out of patient care and focus more on leadership and training—in other words, developing people. It was a big change for the office. I had been extremely busy doing cosmetic surgery and other physical tasks, and then I went to seeing patients only about four to eight hours a week and doing only selected surgeries. This new schedule allowed me to hyper-focus on doing the things I love beyond patient care. I read two business books per week to hyper-drive my education so that I can effectively lead this large, thriving medical practice. I conduct numerous team meetings each week. I meet with key employees on a regular basis. In the past, I had one-on-one meetings only if a team member or doctor was creating a problem, but now I meet with them to help them grow within the organization. All of this allows me the time and luxury to write a couple of books on topics I've been thinking about and studying for quite some time.

In order to be able to write this book, I read abstracts from about 4,000 clinical studies, then narrowed that information down to hundreds of peer-reviewed medical journals. To give you an idea of how intense this onslaught of reading was, I used Adobe Acrobat to count the number of pages I had to read or review. The total? About 7,400 pages of the narrowed-down journals. I printed out about 2,000 of the more pertinent pages for easier reference. This book summarizes the findings of this medical information and clinical studies—it is, in effect, a review of the contemporaneous medical literature concerning the effects of and treatment for low testosterone. Before I changed my daily routine, there was no way I could have accomplished this task—even though I loved doing surgeries and seeing patients, spending most of my time doing that prevented me from doing the other things I love to do.

Not only is the timing of this book perfect but also this information is sorely needed, because the FDA's guidelines have made doctors and patients afraid of using testosterone to treat low testosterone even though it is extremely safe—and clearly beneficial!—to do so.

# 1

# How the FDA Changed
# Everything

## January 31, 2014: The date everything changed

Prior to 2014, there was a 10-year progression of patients and physicians becoming enlightened about the value of TRT (testosterone replacement therapy) for protecting men's health and well-being. Rates of testosterone replacement for men with "Low T" rose 400% between 2003 and 2013, notably in men younger than 45, an age at which low testosterone is not rare but yet is frequently overlooked. That trend came to an abrupt end when the FDA issued a safety announcement to doctors and manufacturers and advised them to alter their prescribing habits. The agency made this announcement based largely on two published clinical papers that later turned out to be flawed.

What prompted the FDA to do this? In January 2014, the FDA was advised—by its own advisory panel—that testosterone could cause an increase in heart attacks and strokes. This was in direct contradiction of more than 30 years of research that showed that TRT delivered cardiovascular *protection*. Hundreds of clinical studies showed benefits, but then two flawed papers showed increased risks . . . and the FDA went with those two (most recent) studies.

In a nutshell, the main study that showed cardiovascular risk in conjunction with TRT was published on November 6, 2013, in the *Journal of the American Medical Association*, one of the most respected medical journals in the United States. Even though 26 medical societies called for a retraction of the study because it had obvious and serious flaws, this paper—and the FDA's immediate reaction to it—became big news.

After the study was published, after the story was picked up by all the news media outlets, and after doctors were advised of the new FDA recommendations, that same study was found to have a serious flaw: *the researchers had incorrectly calculated the statistics.* There was actually **a 50% reduction in cardiovascular** events in men taking testosterone. Not an increased risk, no—rather, a substantial reduction. This correction was published, but the corrected results did not make it into the news and did not sway the FDA. The agency established, maintained, and has even since updated its guidelines based on false information. (The corrections were also published in the *Journal of the American Medical Association,* so they were hardly a secret.)

The FDA went on to further advise doctors and drug companies to prescribe testosterone replacement for men only when the identifiable causes of low testosterone could be linked to clearly documented organ damage (brain or testicles) or an extra female chromosome. Natural declines in testosterone due to aging—which had been successfully treated for decades—should no longer be a reason to treat men, the FDA said. This is one of the most absurd statements to ever come out of the FDA.

The FDA specifically acknowledged that most prescriptions for TRT were being given to men with low testosterone that resulted from aging. They made it clear that if a specific medical cause besides aging was *not* found, then men should be left to suffer from the consequences of low testosterone. We'll talk more about this shortly, but, essentially, untreated low testosterone in men leads to more heart attacks and strokes, worsened blood pressure, worsened cholesterol, weight gain, metabolic syndrome, and diabetes. This plus a myriad of other "normal aging" diseases can occur in men with low testosterone if they're left untreated.

On September 17, 2014, the FDA reconvened an advisory panel to review the current literature at the time, and the panel found that the FDA's prior claim that TRT could be associated with an increased incidence of cardiovascular events was flawed, noting that there was insufficient evidence to support this prior claim. However, on March 3, 2015, the FDA ignored its own advisory committee's findings and reissued its previous statements and claims. The official FDA stance was that its ruling would remain in place and that the agency would continue to require that testosterone products be labeled with this false warning.

This seems ridiculous, doesn't it? Go to the FDA's website and look at the article titled "FDA Drug Safety Communication: FDA cautions about using testosterone products for low testosterone due to aging" (www.fda.gov/Drugs/DrugSafety/ucm436259.htm) and read it for yourself. One statement in particular stands out: "The FDA has become aware that testosterone is being used extensively in attempts to relieve symptoms in men who have low testosterone for no apparent reason other than aging." (This would be the equivalent of saying "the FDA has become aware that doctors are treating cancer when no cause other than aging has been identified" or "the FDA has become aware that doctors are treating heart attacks in men with normal cholesterol who aren't smokers.")

The article goes on to say that "the benefit and safety of these medications have not been established for the treatment of low testosterone levels due to aging, even if a man's symptoms seem related to low testosterone."

This statement flies in the face of thousands of clinical papers addressing just that issue. Let's look at a specific example of just how absurd this is.

*Six-year study of more than 80,000 men with low testosterone*
A study done by the US Department of Veterans Affairs in conjunction with the Division of Cardiovascular Research in Kansas City, Missouri, in 2015 challenged the FDA's claim about testosterone causing an increase in the incidence of cardiovascular events. The results of this study were published in an article titled "Normalization of Testosterone Level is Associated with Reduced Incidence of Myocardial Infarction and Mortality in Men."

Now, there have been countless studies suggesting this same thing, but this particular study was a little different—the researchers did the study specifically to address the FDA's seemingly ridiculous concerns as well as to see whether the study that had ultimately wound up doing a flip-flop in the *Journal of the American Medical Association* made any sense at all.

In this study, researchers evaluated 83,010 aging male veterans with documented low testosterone levels who were offered testosterone replacement. (Mind you, this was the biggest study ever done regarding this issue.) They began by dividing the men into three groups. All had age-related low testosterone and hence would not qualify for testosterone replacement under the FDA's ridiculous guidelines. These men were offered TRT because it is the standard of care even though the FDA currently does not support it. The average age of the men was 66, and the length of time they were followed was a little over six years.

The test groups were as follows:

- Group 1: Low testosterone and given testosterone replacement to normalize testosterone levels (as confirmed by a blood test showing normal testosterone).

- Group 2: Low testosterone and given very little testosterone; did not normalize testosterone levels.

- Group 3: Low testosterone and no testosterone replacement.

The outcomes?

*Group 1*: The men with low testosterone who were given adequate replacement were about 50% less likely to have a heart attack or stroke or to die than the men who were not treated. That's right—they had a significant reduction in heart attacks and strokes. They were also 50% less likely to die from *any* cause—regardless of the cause, their rate of mortality during the subsequent six years after initiating testosterone replacement was *half* that of the men who were not treated with testosterone replacement. In short, they lived longer.

*Group 2*: The men with low testosterone and inadequate replacement were less likely to survive the subsequent six years than the men who were adequately treated with TRT, but the men in group 2 were more likely to survive than the untreated men in group 3.

*Group 3*: The men who received no TRT whatsoever had the highest mortality—they were twice as likely to die as the men who were treated with TRT.

To sum up, more than 80,000 men who were studied over a long period of time were more likely to survive and have fewer heart attacks and strokes if they received TRT than if they didn't receive testosterone replacement (or were not given enough). Yet the FDA guidelines state that these men should not have been treated in the first place because their low testosterone levels were related to getting older.

It makes me sick to look at the FDA bias. This is a government agency that is supposed to be protecting us! But instead, it looks for the no. *"No, doctor, you should not replace low testosterone in men to improve quality of life, increase sexuality, provide cardiovascular protection, bolster mood, help lose or maintain weight, and preserve muscle and bone mass . . . at least, not unless you can prove there is something wrong with their brain or testicles due to injury or cancer."*

Decades of research have demonstrated the safety and efficacy of testosterone replacement in men with low testosterone—and evidence gathered in 2015 has shown a 50% reduction in all-cause mortality in older men—yet the FDA does not want us to use it. I cannot think of a class of drugs that matches natural hormone replacement in terms of safety and benefit ratios that is also supported by one of the largest drug safety studies ever conducted. More than 80,000 men in a study is pretty large! Most clinical studies require far fewer participants than that in order for the FDA to approve a totally new drug with no historical track record.

The FDA's stance on not treating men with low testosterone because it is part of normal aging is by far the most biased statement I have ever encountered from the FDA. Heart disease, cancer, hypertension, obesity, diabetes, strokes, hip fractures, hair loss, erectile dysfunction, prostate enlargement, memory loss, Parkinson's, etc., can all be lumped into the "normal aging" category, yet the FDA has graced us with the permission to treat these conditions.

## The FDA doesn't regulate the practice of medicine

To be clear, the FDA does not regulate the practice of medicine—rather, it regulates the development and marketing of drugs (and food). This means that while the FDA can make recommendations to physicians, it does not have oversight as to how the doctor-patient relationship works.

Also, the FDA ruling that testosterone for men be incorrectly labeled as possibly causing increased heart attacks and strokes does not establish the claim as a fact. The FDA's own advisory panel acknowledged that this statement by the FDA lacked a scientific basis. The FDA's insistence that TRT be restricted only to men who have organ damage or female characteristics from an extra female chromosome does not prevent doctors and patients from making treatment decisions based on sound scientific evidence.

Testosterone was initially FDA approved in the 1950s for men who had damage to their testicles or brain or had an extra female chromosome—and only if their testosterone was severely low and the men demonstrated shrinkage of their muscles and genitalia. Since that is what the FDA approved it for in the '50s, it has maintained its position in spite of countless research studies since that time.

This is not the first blunder by the FDA—historically, this federal agency has gone after manufacturers in similarly absurd ways. Prior to the time Congress passed the Nutrition Labeling and Education Act in 1990, food manufacturers would face fines if they stated important consumer information such as saturated fat, carbohydrate, fiber, and protein content on food labels. In the '70s, the FDA actually considered it illegal to print the nutrition facts on the labels of foods! Its claim was that by labeling foods with nutrition information, consumers might have selected which foods they wanted to consume on the basis of those labels. The FDA was correct in that Americans want to know nutrition information, but the FDA also felt that individuals should not choose which foods they should consume based on nutrition facts.

In 1992, when the Centers for Disease Control (CDC) recommended that women of childbearing age take folic acid—a vitamin to prevent birth defects—the FDA warned manufacturers that if they announced this CDC recommendation to women, they would be prosecuted. Congress fixed this overreach of the FDA in 1994 with the Dietary Supplement Health and Education Act, which President Clinton signed into law. It states:

> *After several years of intense efforts, manufacturers, experts in nutrition, and legislators, acting in a conscientious alliance with consumers at the grassroots level, have moved successfully to bring common sense to the treatment of dietary supplements under regulation and law.* (health.gov/dietsupp/ch1.htm)

The FDA has also removed safe and proven devices from the market, citing the possibility of phantom risks that ultimately wind up requiring patients and physicians to use inferior products. I do cosmetic surgeries, so I am aware of how the FDA drove a large breast implant manufacturer out of business by suspending its production of silicone implants. Much of the FDA's decision was based on the statements of a disgruntled engineering lawyer from the same manufacturing company as well as several women's opinions of the device. After years of additional research (it was the most thoroughly researched medical device in history), the implants were found to be not only safe but also superior to the devices that the FDA had allowed to remain on the market. They are available again and have the FDA's blessing . . . except that a different manufacturer now makes them, because the government's actions bankrupted the first manufacturer.

All that said, this book is not going to be me "exposing" or complaining about the FDA. The people there do good work, but like any government agency, the FDA is influenced by politics and run by people. The agency is not infallible. And by nature, when the FDA makes new claims or establishes new guidelines, the agency is very hesitant to remove them.

Unfortunately, though, now that the FDA has labeled testosterone with misguided information, it is indirectly suppressing the normal practice of medicine. Doctors who prescribe testosterone are aware of the new labeling, and even though they may know that the labels are inaccurate if they have been following the current medical literature, there is a fear of litigation. For example, the study that the FDA cited in its requirement that manufacturers label testosterone with "may cause increase of heart attacks" was corrected to show that there was a 50% *reduction* in heart attack deaths in the men who participated in the study. Other government-sponsored studies have demonstrated not only fewer heart attack and stroke deaths but also significantly fewer deaths overall in the study participants.

Still, we know that men can nevertheless have heart attacks and strokes and that they'll eventually die. Testosterone has been shown to be protective, yes, but it will not result in immortality. So when an incident does occur, there may be litigation because—as a lawyer might say—"The FDA said this could happen, and he [the litigating patient] shouldn't have had TRT in the first place because there was nothing wrong with his brain or testicles."

## Breaking down myths and fears

The purpose of this book is to allow you to have a better understanding of the impact an essential hormone—testosterone—has, how it impacts your future health, and what you can do if your natural production starts to dip. There is much confusion in the medical community over many aspects of TRT:

- At what level treatment should be initiated (based on blood work)

- How to monitor or dose

- The ideal route of administration

Health benefits derived from TRT improve the risks associated with low testosterone, which we will break down into:

- Weight gain, metabolic syndrome, and diabetes

- Cardiovascular disease

- Sexual health, desire, and erectile function

- Bone and muscle decline

- Prostate health

- Brain and mood

In addition to discussing the benefits of TRT, I will address any potential risks associated with testosterone replacement and how to avoid them, as well as specifically address the many myths that have surfaced about the use of testosterone in men.

The entire focus of this book is to discuss true medical replacement of testosterone. A lot of confusion and crossover bias occurs when people think of testosterone as an anabolic steroid. While it's true that bodybuilders use anabolic steroids (which are analogs to testosterone) to create unnatural physiques, we are not going to review the risks associated with those drugs. That is not part of this book and certainly not part of our practice. We replace testosterone in men to promote health and well-being and to improve the quality of life for those who have diminished production of testosterone.

The things we will talk about in conjunction with TRT are, for the most part, "off-label" issues. Off-label means using a drug that is approved by the FDA but is not approved for that specific use. This is an allowed, legal, and common practice in health care, as outlined on the FDA's website in the article titled "Understanding Unapproved Use of Approved Drugs 'Off-Label'" (www.fda.gov/ForPatients/Other/OffLabel/ucm20041767.htm).

Testosterone has been approved for men since the 1950s, yet on March 31, 2015, the FDA updated its 2014 recommendation with a drug safety communication, stating:

> [03-03-2015] The U.S. Food and Drug Administration (FDA) cautions that prescription testosterone products are approved only for men who have low testosterone levels caused by certain medical conditions. The benefit and safety of these medications have not been established for the treatment of low testosterone levels due to aging, even if a man's symptoms seem related to low testosterone. We are requiring that the manufacturers of all approved prescription testosterone products change their labeling to clarify the approved uses of these medications. We are also requiring these manufacturers to add information to the labeling about a possible

*increased risk of heart attacks and strokes in patients taking testosterone. Healthcare professionals should prescribe testosterone therapy only for men with low testosterone levels caused by certain medical conditions and confirmed by laboratory tests . . . .*

*Testosterone is FDA-approved as replacement therapy only for men who have low testosterone levels due to disorders of the testicles, pituitary gland, or brain that cause a condition called hypogonadism. Examples of these disorders include failure of the testicles to produce testosterone because of genetic problems, or damage from chemotherapy or infection. However, the FDA has become aware that testosterone is being used extensively in attempts to relieve symptoms in men who have low testosterone for no apparent reason other than aging.* (www.fda.gov/Drugs/DrugSafety/ucm436259.htm)

During the course of this book, I will lay out a summary of the most relevant scientific data disputing the FDA communications. If there is information supporting the FDA, I will explain that as well, but as I've said previously, remember that the FDA made the new guidelines based on faulty studies that have largely been disproven and/or found lacking in sound scientific principles. And the criteria it suggests today (only use for brain and testicle disorder patients) are based on the initial research done more than 60 years ago. This is not my opinion alone; what we will be talking about has been and still is supported by numerous medical societies and expert panels.

*The position of the American Association of Clinical Endocrinologists*
"The American Association of Clinical Endocrinologists and the American College of Endocrinology Position Statement on the Association of Testosterone and Cardiovascular Risk" was published

in *Endocrine Practice* on September 2015 (vol. 9). Endocrinologists are physicians who specialize in treating hormone disorders.

In the association's position paper, it makes five key points that are at odds with the seemingly reckless FDA guidelines:

1.  Studies show that low testosterone is associated with increased cardiovascular events and all-cause mortality, an association that gets stronger as men age.

2.  Testosterone replacement therapy favorably changes cardiovascular risk factors by decreasing fat mass and decreasing insulin resistance; also, TRT can reverse metabolic syndrome, which is the precursor to diabetes.

3.  The two studies the FDA used to make the recommendations for labeling testosterone with warnings about heart attacks and strokes had major flaws, thus precluding any meaningful conclusions.

4.  The FDA conclusion that there may be an association between testosterone replacement and heart cardiovascular disease is incorrect.

5.  The FDA recommendation that testosterone be replaced only in men with "disorders of the testicles, pituitary gland [part of the brain that signals the testicles to make testosterone], or brain" is just plain wrong. The association goes on to agree that testing may be indicated to check for those conditions, but if there is nothing wrong with a man's brain or testicles, then additional testing and treating for signs and symptoms of low testosterone should guide the decision-making process.

In its position paper, the association also reviews the flawed studies that the FDA misguidedly used to make its recommendations. While the FDA finds that there is an increase in cardiovascular events, the data are misrepresented—in the study, there are 50% *fewer* cardiovascular events in the men taking testosterone versus those not taking it. In addition, two corrections were published in the same medical journal that had originally published the flawed results. The authors of the original study admitted to a series of data errors in the study and published that admission, but the FDA nonetheless kept its recommendations in place. Twenty-nine medical societies have called for a retraction of the article based on the misinformation it contains and the weight the FDA gave the flawed study.

The association goes on to refute the FDA's position on prescribing testosterone only to men with disorders of the brain or testicles—the association says that ruling is very shortsighted. It identifies the wealth of scientific data supporting the use of TRT in men with low testosterone and states that the FDA had no valid basis for its guidelines. The association also cites the numerous conditions that occur when a man's testosterone declines, as well as the benefits he sees when it is replaced.

*International Expert Consensus Resolutions*

"Fundamental Concepts Regarding Testosterone Deficiency and Treatment: International Expert Consensus Resolutions" was published in the *Mayo Clinic Proceedings* in July 2016 (91[7]: 881–896) and is the outcome of recommendations from an international panel. The reason for the consensus meeting was to directly respond to the way the FDA had erroneously made new guidelines for testosterone replacement based on two flawed studies. King's College London and the International Society for the Study of the Aging Male supported the resolutions. The authors of the resolutions acknowledged the widespread and intense media attention the two studies received, the errors of the studies, and the confusion that accrued as the result of false data circulating within the medical community.

The panel consisted of a broad range of medical professionals and experts representing various specialties and coming from 11 countries on four continents. The members of the panel focused on scientific data demonstrating the highest content of quality and vowed to make their resolutions only if all experts unanimously gave their approval to do so.

They noted that since the 1990s, clinical research showed that men with low testosterone would benefit from testosterone replacement. In the 2000s, the use of TRT became more widespread as testosterone formulations became easier to use. All of the research leading to this point showed that medical professionals should be evaluating men for low testosterone and treating them when appropriate. This came to a halt when the FDA made new guidelines based on faulty information, guidelines that the agency initiated *within two days* of the publication of the faulty study. It's true that testosterone replacement associated with illicit athletic performance enhancement has long been controversial; this may be part of the reason for the FDA's unusual bias and lightning-fast action. Was the FDA looking for a weakness in testosterone's "armor" of supportive research, so to speak, and a reason to put a stop to the use of this drug despite the fact that it has been shown not only to be safe but *also* to reduce rates of all-cause mortality?

## Summary of the International Expert Consensus Resolutions

1. *Testosterone deficiency is a well-established, significant medical condition that negatively affects male sexuality, reproduction, general health, and quality of life.*

**Low testosterone predicts increased risk of obesity and metabolic syndrome. Low testosterone is associated with increased death rate and more cardiovascular events and deaths. It also is associated with a lower quality of life and poorer general health.**

Testosterone deficiency has been identified as an important medical condition since 1940. Testosterone affects virtually all organ systems, so a deficiency negatively impacts virtually all organ systems. The panel referred to testosterone deficiency as a major medical condition that negatively impacts general health and quality of life as well as men's sexuality.

The panel identified the clear links between low testosterone and increased fat mass, loss of muscle mass, increased insulin resistance (which leads to diabetes), worsened cardiac lipids, increased cardiovascular risk, increased depression, less enjoyment of life, decreased orgasmic function, decreased erectile function, and fatigue.

The panel went on to say that the general medical community needs to have a greater awareness of the effects of untreated low testosterone in men.

2.  *The symptoms and signs of testosterone deficiency occur as a result of low levels of testosterone and [men] may benefit from treatment regardless of whether there is an identified underlying etiology.*

**This means that regardless of the cause of low testosterone, men benefit from testosterone replacement. The panel additionally pointed out that it is inappropriate to withhold testosterone replacement when a cause cannot be identified, which is usually the case.**

The symptoms of low testosterone are due to the actual low testosterone, not the underlying cause. The FDA approves testosterone replacement for men with organ damage and men who have rare genetic conditions wherein they have an extra female chromosome and don't fully develop male characteristics. In addition, there are men who have had brain tumors or trauma; there are also men with testicular disease such as injury or failure to descend. Despite these conditions, however, the vast majority of men with low testosterone acquire it as they age.

Large clinical studies have shown that men with low testosterone benefit from testosterone replacement regardless of the cause—the FDA guidelines are baseless and are not based on any current scientific evidence. This is the very nature of bias. Studies carried out more than 60 years ago focused on testosterone replacement in men with disorders of the brain and testicles; then later (starting in the late '90s), studies were done on men with low testosterone in general with normal brains and testicles. But nonetheless, in 2014, the FDA revised testosterone replacement guidelines based on two flawed studies as well as very dated scientific evidence done decades ago on men without testicles.

In its expert consensus paper, the panel says, "Thus, there appears to be no scientific basis to recommend restricting testosterone therapy only to men with an identified underlying etiology, although there is value in attempting to identify such conditions when possible."

3.   *Testosterone is a global health problem.*

**Here, the panel explained the enormous costs (in the hundreds of billions of dollars in the United States alone) associated with untreated low testosterone.**

How often testosterone deficiency is identified is largely based on a cutoff laboratory value (which can be fairly arbitrary). We will talk a lot about this later, but, in general, a lab test is used to identify low testosterone. Where you draw the line (is the testosterone a little low or a *lot* low?) will determine whether a man will be told that he has low testosterone. A diagnosis of low testosterone is probably given to about 30% of middle-aged men; many more senior men probably are told they have low testosterone.

Most of the increased healthcare costs of untreated low testosterone are related to conditions associated with obesity, such as diabetes, heart attacks, and premature deaths. It's estimated that even when using a low cutoff value to identify low testosterone in men (i.e., 300 ng/dL or 10.4 nmol/L), only 12% of men with low testosterone in the United States are treated with testosterone despite having access to health care. And this cutoff value doesn't just indicate low testosterone; it indicates *very* low testosterone.

Unfortunately, despite the testing that's being done, doctors are reluctant to prescribe testosterone even though it will lead to lower rates of obesity, diabetes, heart attacks, strokes, and all-cause mortality and will simultaneously lead to various improvements in quality of life. Only 12% of men with low testosterone are treated, and these men register as having "very low" testosterone. The vast majority of men with low or very low testosterone are not treated.

4.   *Testosterone therapy for men with testosterone deficiency is effective, rational, and evidence based.*

**Testosterone replacement definitively benefits sexual desire and erectile and orgasmic function. Testosterone replacement increases lean body mass and decreases fat mass. Testosterone replacement improves bone mineral density, mood, and energy.**

In numerous studies, testosterone replacement has positive effects on multiple organ systems when treating men with low testosterone. Diet and exercise alone are beneficial for improving weight, muscle mass, and energy, but in men with low testosterone, those benefits are magnified when men also receive testosterone replacement (compared to those who don't).

In its position paper, the panel concludes that "testosterone therapy for men with TD (testosterone deficiency) is effective, rational, and evidence-based when used to treat issues related to sexual function and body composition."

5.   *There is no testosterone concentration threshold that reliably distinguishes those who will respond to treatment from those who will not.*

**In this resolution, the panel unanimously agreed that the absolute value of testosterone in a blood test is not a predictor of or an accurate way to diagnose testosterone deficiency. Symptoms of testosterone deficiency as well as the overall clinical picture should be combined with a laboratory evaluation to assist in the decision of whether to prescribe testosterone replacement.** The best and most appropriate way to determine whether a symptomatic man will respond to TRT is to assess his response to TRT.

A quick note: because testosterone is commonly measured in either nanograms per decaliter or nanomoles per liter, I will use both here. (Our lab uses only nanograms per decaliter [ng/dL], so I'll use this measurement first.) Both the expert panel members who wrote the resolutions and the Endocrine Society agreed that the absolute level of testosterone measured in the blood does not reliably predict who will benefit from testosterone replacement, but here are some target levels.

- At 430 ng/dL (15 nmol/L), many men will become symptomatic. This may include decreased energy and sexual desire.

- At 350 ng/dL (12 nmol/L), men will tend to put on visceral fat (belly fat); below 290 ng/dL (10 nmol/L), men will trend toward diabetes.

- Below 230 ng/dL (8 nmol/L), men will trend toward erectile dysfunction as well as develop other "age-related" conditions.

If you search your insurance plan guidelines for what qualifies men to receive testosterone replacement, this is what you will generally find:

*The testosterone must be very low (generally in the range of 250–300 ng/dL or below) and must be caused by a specific medical condition, such as organ damage or a genetic defect that results in boys not fully developing into men, for it to meet FDA approval and in some cases for your insurance to cover it. You will likely be required to test testosterone levels twice a day on separate days at the time when testosterone production is the highest, which is between 8 and 10 a.m. The plan will then select the higher set of values to determine that you are not eligible. If somehow you are eligible for TRT, the plan will only cover receiving the amount that will take you from very low testosterone to low testosterone.*

The bottom line is this: don't let the insurance industry or a biased government agency make decisions for your body. Medical decisions are between you and your doctor. Fortunately, testosterone replacement isn't particularly expensive, which is good since you will most likely have to pay for it yourself.

In the consensus resolutions, the panel acknowledged that the total testosterone concentration should be measured, yes, but that the absolute level cannot predict who will and who will not benefit from testosterone replacement. Different individuals have different thresholds where symptoms will occur, for example. There are many and varied factors that influence the effect of testosterone on our organ systems that just can't be measured.

The panel ended its resolutions with this statement:

> *The diagnosis of TD (testosterone deficiency) should include [an] assessment of the entire clinical presentation, aided by biochemical tests (lab tests). Rigid application of a uniform total testosterone concentration threshold for all individuals as the primary instrument to diagnose TD lacks scientific foundation and is discouraged.*

6. *There is no scientific basis for any age-specific recommendations against using testosterone therapy.*

**Regardless of age, men respond to testosterone therapy when they have low testosterone. Age is not a predictor of testosterone deficiency; in addition, other factors such as genetics, overall health, and obesity play roles in testosterone deficiency.**

**Avoiding testosterone replacement in elderly men is as illogical as not treating other conditions—high blood pressure, heart**

**disease, arthritis, cancer—because they also occur more frequently "as men age."**

The FDA's statement that "the FDA has become aware that testosterone is being used extensively in attempts to relieve symptoms in men who have low testosterone for no apparent reason other than aging" flies in the face of scientific evidence that demonstrates that age is not a consistent predictor of declining testosterone and that other comorbidities such as weight gain are associated more closely with low testosterone than they are with aging.

The FDA also notes that total testosterone levels may remain stable and that other factors may reduce free testosterone availability. Free testosterone is not bound to inactive proteins and is available for your body to use.

Regardless of cause, a similar beneficial clinical response to testosterone replacement occurs in both younger and older men with low testosterone. The panel noted that since life expectancy is increased with testosterone replacement, there is "no justification to recommend restricting testosterone therapy based on age."

Yes, low testosterone is more common in older men . . . just like hypertension, diabetes, heart disease, and cancer are more common in older men. Although it does increase the likelihood of low testosterone, age does not assure low testosterone (or other diseases, for that matter).

Again, not treating medical conditions just because they occur more frequently as we age is just about the most absurd concept in health care. The FDA does not recommend restricting any other drug because of its connection to aging—hypertension, diabetes, cancer, cardiovascular disease, and obesity all occur more frequently as men age, and the FDA has not ruled that treatments be withheld for those conditions because they are associated with "normal aging."

7.  *The evidence does not support an increased risk of cardiovascular events with testosterone therapy.*

**Numerous clinical studies have shown benefits in testosterone replacement in men who have heart disease, such as a greater capacity for exercise and less chest pain. On the other hand, low testosterone is known to be associated with increased heart disease, obesity, and increased all-cause mortality.**

**Two studies done between 2011 and 2013 showed a possible link between testosterone replacement and decreased cardiovascular health. These studies were flawed (and later corrected), and their outcomes were not supported in numerous other studies (which showed positive outcomes in conjunction with testosterone replacement).**

These two negative studies were notably the lone sailors in an ocean of studies showing positive outcomes. More than 100 prior clinical studies show cardiovascular benefit with testosterone replacement; they also demonstrate the clear association between low testosterone and increased cardiovascular disease. The flawed study done in 2013 (which was later corrected) showed an increase in cardiovascular events after initiating testosterone therapy. Once the study's flaws were pointed out and corrected, the data actually showed a 50% reduction in cardiovascular events, but that correction didn't make the news. Not surprisingly, the flawed study was widely publicized—news is more attention grabbing when it flies in the face of what we have known and believed for years.

CNN published a blog that popularized the potential risks that were later found to be based on false interpretations of the data. The comments on the blog left by the readers are interesting—even the nonscientific readers who read the blog picked up on the fact that something didn't sound right, but still, the FDA bought into it. You can read the blog and comments here: *www.thechart.blogs.cnn.com/2013/11/05/testosterone-treatment-could-be-dangerous-to-the-heart/.*

The *New York Times* also covered the flawed study, with numerous stories and blogs dedicated to it. In a blog post titled "Overselling Testosterone Dangerously" (on its opinion page), it states:

*A large study has found substantial risks in prescribing testosterone to middle-aged and older men for a variety of ailments. One part of the study found that testosterone doubled the risk of cardiovascular disease in more than 7,000 men who were 65 years old or older, essentially confirming findings in previous studies. The other part found that testosterone almost tripled the risk of heart attacks in a group of more than 48,000 middle-aged men with previous histories of heart disease. The harm in both cases occurred within 90 days of receiving the prescription.*

The full post is available at www.nytimes.com/2014/02/05/opinion/overselling-testosterone-dangerously.html?_r=0.

It turns out that the referenced study contains errors and is not meaningful. Also, the statement "essentially confirming findings in previous studies" is incorrect—hundreds of studies show either a neutral or protective effect of testosterone on cardiovascular disease. The largest study ever done on testosterone that included more than 80,000 men who were followed for about six years (more people followed for a longer time results in higher-quality and more meaningful conclusions) showed significantly fewer (50%) cardiovascular events and significantly lower rates of all-cause mortality. In the 80,000-man, long-term study, the men treated with testosterone replacement were subsequently half as likely to die as the men with low testosterone who were not offered replacement.

But because news needs to be sensational and critical and have a villain, the *New York Times* peppered the study up even more, stating:

*The reason seems clear. Drug companies have shamelessly pushed the notion, to doctors and to the public, that their testosterone-boosting product can overcome a supposed disease called "Low T," which is characterized by feelings of fatigue, loss of sexual drive, depressed moods, an increase in body fat and decrease in muscle strength, among other symptoms.*

*Incredibly, AbbVie, which makes the market-leading testosterone gel known as AndroGel, lists "Low T" as one of five important health risks facing men, along with high cholesterol, high blood pressure, high blood sugar and high levels of prostate-specific antigen that may indicate prostate cancer.*

All of this gives the *New York Times* a good article, but it based its statements on false information. In fact, the "villain" in this case—the drug company—was offering men something that could save their lives. AbbVie was absolutely correct in its marketing of the product.

If you Google all the major media outlets for stories about that study, you'll find that the big headline at the time was that evil "Big Pharma" was selling a dangerous drug. Then shortly thereafter, research came out that the data the study was based on were false and that there was actually a substantial decrease in cardiovascular events, *not* an increase . . . but there wasn't much buzz about the correction. When CNN and the *New York Times* were proven to be wrong, they simply did not talk about it.

The *Wall Street Journal*, which typically isn't as splashy as some of the other media outlets are, did write articles about the updated information: on July 4, 2014, it ran "Testosterone Use Doesn't Increase Heart Risk, Study Finds Findings Run Counter to Earlier Research," by Melinda Beck (www.wsj.com/articles/testosterone-use-doesnt-increase-heart-risk-study-finds-1404423497). The article acknowledges that the men the study followed had a substantially lower risk of heart attacks if they took testosterone replacement. The article discusses the controversial study that the FDA used as a basis for so much of its decision to put a warning label on and restrict use of testosterone prescriptions. The article goes on to say that "critics have attacked the study's methodology for, among other things, including over 100 women among the 1,132 subjects studied. Over 25 international medical groups have demanded that *JAMA* retract the article. *JAMA* has declined to do so."

*JAMA* is the *Journal of the American Medical Association*, an international peer-reviewed journal. It didn't retract the article, but it did publish the corrections. These corrections were barely noticed and were not picked up by media outlets the way the initial incorrect study was. In the cardiovascular section of this book, we will talk about the flawed data a bit more.

8.  *The evidence does not support an increased risk of prostate cancer with testosterone therapy.*

**Testosterone therapy does not increase the risk of prostate cancer. Men previously treated for prostate cancer who are in remission can safely use testosterone replacement.**

We'll cover the testosterone-and-prostate-cancer link in depth in another section. Just briefly, many doctors believe that testosterone replacement can potentially increase a man's risk for prostate cancer. It turns out that is a myth. The root of that myth is the known fact that some types of prostate cancer respond to a treatment that involves lowering a man's testosterone level to almost zero. Those protocols led to the belief that "if we turn testosterone off and some prostate cancers get better, then maybe it is better to have low testosterone to avoid prostate cancer altogether." That could have been a legitimate hypothesis, but it turns out to be wrong. Why? There is a threshold effect. If the testosterone level is between 200 and 1,000 ng/dL, yes, certain prostate cancers can continue to grow, in which case turning testosterone down to zero will slow the cancer or completely stop it from growing. However, raising the testosterone level does not make the cancer more likely to grow faster. In fact, certain types of prostate cancer are specifically treated by *raising* the testosterone level, so it isn't accurate to categorically state that low testosterone prevents all types of prostate cancer.

The bottom line is this: testosterone replacement does not increase a man's risk of developing prostate cancer. PERIOD!

9. *The evidence supports major research initiatives to explore possible benefits of testosterone therapy for cardiovascular and metabolic diseases, including diabetes.*

**Men with low testosterone who receive testosterone therapy have lower rates of premature mortality than untreated men—in large-scale studies, premature mortality was reduced by half. Men treated with testosterone therapy have fewer cardiovascular events and premature deaths than men who are not treated.**

**Testosterone therapy reliably increases lean body mass and decreases fat mass.**

**Higher levels of testosterone are protective to the heart and cardiovascular system.**

**Low testosterone is associated with type 2 diabetes, metabolic syndrome, obesity, and cardiovascular disease, and long-term studies have shown that treating these conditions with testosterone replacement results in a twofold improvement of survival.**

This particular resolution is crucial. Numerous studies show the benefit of testosterone replacement for men with a testosterone deficiency. These benefits include better quality of life, better mood, increased sexuality, increased vigor, leaner body with less obesity, fewer cases of metabolic syndrome, reduced incidences of diabetes, fewer cardiovascular events, and lower rates of all-cause mortality. To be clear, testosterone replacement does not *prevent* death, diabetes, and obesity, but it does lower the risk of and offers protection against those outcomes. We are all still going to die—and men with testosterone replacement still may have heart attacks—but deaths and heart attacks are likely to occur later in life for men treated with testosterone compared to men with low testosterone who are left untreated.

To recap, testosterone replacement had been more understood and more commonly accepted by doctors and their patients since the late '90s, with about a fourfold increase in prescriptions starting in the year 2000. Doctors were becoming educated, consumers were becoming more aware, and testosterone replacement was generally more acceptable—it was a pattern similar to the one we saw with respect to treating high blood pressure decades ago. (At the time, we didn't know what caused hypertension, but when we understood the impact of treating it, treatments became more accepted and widespread.) The scientific data were consistent: testosterone replacement was repeatedly shown to lead to better health in men with low testosterone.

## When bad news is flawed news

Everything changed between 2013 and 2015 when two flawed articles were published that—for the first time—suggested an increased cardiovascular risk in men who took testosterone replacement. Even though the studies were found to be flawed and the corrected data published months later actually showed a 50% *reduction* in cardiovascular risk, the die had already been cast. The FDA immediately reacted and published new rules for testosterone replacement, in doing so ignoring decades of research and the opposition of 29 medical societies and countless experts and instead aligning itself with the plaintiff lawyers, the popular media, and the anti-pharmaceutical groups.

Despite the mountains of evidence to the contrary, the FDA changed its guidelines to state that testosterone replacement may increase the risk of cardiovascular disease and stroke, thus restricting TRT to men who have organ damage and/or an extra female chromosome. This remains the FDA's stance even though the since-corrected studies the agency had based its findings on actually show *reduced* rates of cardiovascular risk.

The FDA is also critical of doctors who are treating their patients with the best interests of the patient in mind. The FDA's advisory board is even critical of the very practice of medicine, noting that about 60%

of testosterone prescriptions are written by primary care doctors, not specialists in brain or testicle disorders. The FDA says that calls into question the scope of the practice of family doctors and the relationships between them and their patients.

This whole fiasco is going to do some damage for years to come. Yes, some of the plaintiff lawyers are going to get rich—more than 1,000 suits in federal courts are attempting to get money out of testosterone manufacturers. And some people with large followings who know absolutely nothing about the facts or science are going to get accolades for attacking "Big Pharma." And, yes, when the study first came out, news outlets did get lots of attention with a very controversial—albeit false—story.

But the real victim in this bungle between a small group of researchers and the FDA is potentially you, because there is a very good chance that your doctor is aware of the new FDA rules (doctors should be aware of FDA rules on prescriptions they write). Your doctor may have heard on CNN that the FDA has established new guidelines and that the flawed study showed—at least at first—an increase in heart attacks. But not many doctors may know that the study has been disproven.

The FDA does not regulate the practice of medicine, but its rulings do impact it. How? If the FDA says that testosterone is to be used only for men with specific diseases of the testicles or brain, then manufacturers cannot talk about or promote other uses of testosterone the way they did when they ran their "Low T" campaigns—those promotions came before the FDA's new guidelines came into effect. Also, insurance companies may elect *not* to pay for testosterone replacement in the absence of a damaged brain or damaged testicles. Remember that in order for insurance to pay for testosterone replacement, a man's testosterone level has to be very low and has to be proven through doing repeated blood tests in the morning, when testosterone is at its peak.

If, however, your healthcare practitioner is up-to-date on the current evidence regarding testosterone replacement, he or she will feel comfortable discussing the facts and the risk/benefit ratio with you. In a nutshell, the risks stem from not taking testosterone replacement either when you have low testosterone (or are taking too low of a replacement dosage) or when you are taking too much testosterone. Testosterone levels can be checked with a simple blood test or can even be based on a man's symptoms—it's really not difficult. The benefits will be covered later in this book. In addition, we will carefully examine any other controversies surrounding TRT.

## Making false matters worse

Since the FDA ruling, not only have drug makers had to stop promoting testosterone replacement therapy for men with low testosterone but also they face financial risks due to a variety of lawsuits. That's because the FDA's incorrect claims that testosterone may be associated with an increased risk of cardiovascular events mean that men who are taking testosterone for low testosterone and who subsequently have cardiovascular events now have an opportunity to sue the manufacturers of the testosterone and blame the testosterone replacement for their condition.

We certainly want to protect consumers, but these lawsuits are based on junk science. The fact that testosterone can *reduce* a man's chance of having a cardiovascular event is not the same as testosterone being 100% capable of *preventing* cardiovascular events. Low testosterone leads to cardiovascular events—and testosterone replacement lowers the risk of those events occurring—yet because of the FDA's new guidelines, attorneys and their clients have an opportunity to try to collect money from the manufacturers over illnesses men have had.

On April 6, 2015, *Forbes* ran a story on the 1,340 men who were suing drug manufacturers for various things that happened to them while they were taking testosterone replacement. These men had various medical conditions that are not linked to taking testosterone replacement—interestingly, they had conditions related to low testosterone. In fact, the diseases these men suffered from have specifically been proven *not* to be caused by testosterone replacement even though the plaintiffs' attorneys and news pundits claimed otherwise. The suit alleged that the manufacturers promoted (via direct-to-consumer advertising) the use of their products for a condition that the FDA doesn't recognize as a disease. In effect, the plaintiffs said that "there is no [such thing as] low testosterone." But it's obvious that low testosterone is a condition—it has been studied by scientists for decades. Still, the edge that the plaintiffs and their lawyers had is that the FDA didn't exercise the scientific method rigorously when it changed the guidelines. The *Forbes* article emphasized that angle:

> *"I've seen drugs with inadequate warnings or side effects that they should have disclosed and didn't. But what I haven't seen before is companies inventing a fake disease," says Ron Johnson, a partner at Schachter, Hendy & Johnson in Fort Wright, KY and co-lead counsel on the testosterone litigation cases. [Plaintiff] 'Bob talks about having borderline low testosterone, but there's no such thing as that. It was made up in marketing departments of drug companies. I've never seen this level of disease-mongering.'"*

The lawsuit is extensive—it's more than 128 pages long. But it all comes down to the summary, which states that testosterone prescriptions should be given only to men with drastic hormonal decline caused by injury or disease. "'I'm very hopeful the FDA changes will change prescribing habits,' Johnson says. 'From this point forward, [the lawsuits] are all about getting compensation for these men.'" Read the full article here: www.forbes.com/sites/arleneweintraub/2015/04/06/whats-next-for-the-thousands-of-angry-men-suing-over-testosterone/#7e15639d4042.

Unfortunately, to some degree, the lawyer's wish came true. In the wake of the FDA ruling, doctors often avoid talking about men's testosterone levels and TRT for fear they'll face personal litigation should some health condition with the patient occur. This is a real threat. Even though the science overwhelmingly shows that testosterone replacement does not cause increased risks for the conditions the men are suing about, doctors and manufacturers still have considerable exposure to lawsuits. Part of the reason I'm writing this book is to clarify these risks. How knowledgeable will a jury be when a man is sitting there who had a heart attack and was given a prescription for testosterone a month earlier? Will the jury members be sufficiently educated about studies that show that low testosterone leads to heart attacks and that while testosterone replacement reduces the risk of a heart attack by about 50%, it doesn't reduce it by 100%?

Numerous websites have been created to recruit men who have taken testosterone replacement and suffered various diseases to participate in these lawsuits. The reality is that even when a drug reduces your chances of getting a disease, it does not guarantee against ever getting it. Unfortunately, if the truth is sufficiently twisted, drug companies can be held liable for something that is totally fabricated. When I look at the websites pertaining to the lawsuits, it's clear the lawyers are either manipulating statements and facts or being outright dishonest.

My concern is that as these lawsuits play out, more doctors will be reluctant to prescribe testosterone for fear of personal litigation, which will in turn have a negative impact on men's health. Litigation outcomes are sometimes perceived as establishing fact when in many cases they are just a means to reallocate wealth. Thousands of lawsuits will potentially alter best practices of medicine.

## How this affects your insurance coverage

Health insurance companies have also responded to the FDA rules. Starting in about 2007, sales of testosterone replacement therapy medications went up substantially, and many men received insurance coverage for testosterone replacement. But starting in 2014, insurance companies took notice of the new guidelines, and in many cases, they have stopped paying for testosterone replacement therapy unless certain very strict and specific criteria are met, namely, that men who have very low testosterone must also have organ damage and/or a genetic disorder.

A study in *PLoS* (Public Library of Science) in July 2014 looked at the impact of universal healthcare drug coverage before and after the Ontario government stopped covering the cost of testosterone replacement except in the rare situations of very low testosterone caused by specific brain or testicular disorders. Specifically, the study looked at the rates of testosterone prescriptions written between 1997 and 2012. The turning point was 2006, when Health Canada changed the way it covered testosterone replacement. There was an immediate 26% drop in prescriptions for a few months, followed by a gradual and steady rise.

The lesson? Stopping coverage of TRT does not stop men and their doctors from using TRT. It causes a little downward blip, yes, but the use of TRT keeps on growing. Neither the FDA nor the government nor the plaintiffs' lawyers have stopped men from accessing testosterone replacement. If your testosterone is low, your risk of adverse health outcomes increases. This is your choice. Unfortunately, insurance companies have been a barrier to men receiving appropriate testosterone treatments. Referring again to the *PLoS* study, in 2012, only 11 of 1,000 men in Ontario over the age of 65 were on TRT, and that was up from 3.6 men per 1,000 in 1997. That means that the majority of men with low testosterone are still not being treated.

## The cost of quality of life

Cost is another factor in the ongoing conversation about TRT. In 2014, an article was published in the International Society for Sexual Medicine's *Journal of Sexual Medicine* about weighing the costs of TRT. The title was "Is Testosterone Replacement Therapy in Males with Hypogonadism Cost-effective? An Analysis in Sweden." (Hypogonadism is the medical term for low testosterone.)

The article evaluated the lifelong cost of testosterone supplementation in men who have low testosterone secondary to aging and then evaluated how much TRT costs per additional quality-of-life year. In other words, what is the cost of giving a man another year of his life that is a "quality" year, or a year without substantial disease? The article found that treating low testosterone regardless of its cause is very cost-effective—after all, men with low testosterone have higher incidences of obesity, diabetes, hypertension, and cardiovascular disease. Even though there is an obvious cost associated with purchasing testosterone (perhaps for decades), it still makes economic sense to do so because of the reduced costs associated with the reduced risks of diseases associated with low testosterone.

It's interesting to note that health insurance companies don't want to pay for medications that prevent disease—instead, they are more than willing to pay for management of the disease caused by low testosterone itself.

## The FDA sticks to its guns

On August 20, 2015, after the 80,000-man study confirming that testosterone replacement reduces all-cause mortality and heart attacks, the FDA published a perspective on its prior ruling in the *New England Journal of Medicine*. The agency justified its decision to use a very small number of studies (and didn't acknowledge that those studies were flawed and then went on to be updated and disproven). Amazingly, even after being called out on its bungle, the FDA issued this statement: "To date, there is no definitive evidence that increasing serum testosterone concentrations in these men [referring to men with low testosterone symptoms] is beneficial and safe, and the need to replace testosterone in older men who lack a distinct, well-recognized cause of hypogonadism [low testosterone level] remains debatable."

One of the well-recognized causes cited in the article as being an acceptable reason to use TRT is Klinefelter syndrome, which is a genetic disorder where a male has an extra female chromosome that causes the man to develop into a small person with small genitals, less body hair, female breasts, infertility issues, and little interest in sex. The FDA feels that these men can have testosterone replacement. The agency also finds it acceptable to prescribe testosterone for men who have a pituitary injury. (The pituitary gland is in the center of the brain and provides the necessary connections for the brain to send signals to the organs to make hormones. When it comes to testosterone, the pituitary gland makes luteinizing and follicle-stimulating hormones to stimulate the testicles to make testosterone and sperm.) In the case of this condition, the FDA says that it's okay to use testosterone. The agency also feels that TRT is appropriate when a man has suffered toxic damage to his testicles.

And that's all the FDA allows. If your testosterone levels drop for no apparent reason other than aging, the agency's guidelines state that testosterone replacement should not be used.

To put the FDA's stance into perspective, note that 80% of hypertension (high blood pressure) cases have no known cause. We may eventually figure out what is causing those individual cases, but right now, we just know those patients have to be treated. Yet when it comes to low testosterone, the FDA specifically and repeatedly identifies that if the immediate cause of low testosterone can't be identified, there is no reason to treat it. What's more absurd, the FDA says there is lack of substantial evidence of the use and safety of TRT. Maybe the agency takes this position because of its selective look at the medical literature. The FDA based its decision on too little data that were also flawed data . . . and the agency is sticking with its decision.

There are two major flaws regarding the FDA rules on testosterone replacement. We have talked about why the agency's statement that TRT may increase heart attacks and strokes is incorrect, but its position that testosterone should be used only for men with organ damage or genetic disorders is also ridiculous. This aggressive kind of restriction is not applied to virtually any other aspect of health care. The FDA's assertion either that age-related low testosterone (or what is referred to as "late-onset hypogonadism") doesn't exist or that men won't benefit from testosterone replacement is false as well, as pointed out in a formal position statement by a European medical society, EMAS (European Menopause and Andropause Society). The society states that the diagnosis of low testosterone is justified in men with symptoms and altered laboratory values. Even the FDA says in its position statement that men with low testosterone who receive TRT will benefit by experiencing lower rates of obesity, metabolic syndrome, type 2 diabetes, sexual dysfunction, and osteoporosis. Still, despite these benefits, the FDA says that TRT can be used to reverse low testosterone only if there is something wrong with a man's testicles or his brain or if he has a genetic disorder.

In the *World Journal of Men's Health* in August of 2013, an Italian position paper on TRT was published, titled "Risks and Benefits of Late Onset Hypogonadism Treatment: An Expert Opinion." This came out just prior to the erroneous studies that led the FDA to revise its guidelines for testosterone replacement.

In the article, the authors evaluated the scientific data available at the time and aligned it with myths and fears about replacing testosterone in men with an age-related decline in testosterone. They discussed the associations between untreated low testosterone and chronic conditions such as cardiovascular disease, obesity, bone loss, metabolic syndrome, type 2 diabetes, abnormal lipids, and hypertension. They also noted that testosterone levels are generally not measured or treated when those conditions occur.

The authors pointed out that the medical literature shows that testosterone replacement is without serious adverse events (like me, they did an extensive review of the literature rather than reviewing only a small number of studies the way the FDA did). They also pointed out that the available evidence suggests that testosterone replacement may reduce obesity, improve blood sugar control in patients with diabetes, strengthen bones, and increase muscle mass. Testosterone replacement has also been shown to be beneficial, the authors say, for men with low testosterone who are otherwise healthy as well as for men with low testosterone who simultaneously have debilitating chronic medical conditions.

## In summary

Starting in 2014, testosterone replacement therapy for low testosterone has gone through a bit of a roller coaster.

Over a half century ago, TRT was introduced just for specific conditions where men failed to produce testosterone because of damage to their testicles or brain or because they had a chromosomal abnormality.

Despite the narrow set of original uses for TRT, as physicians became more aware of the symptoms of low testosterone and the diseases that were associated with low testosterone, they began treating more and more men with symptoms of low testosterone (and those with disease resulting from low testosterone) with testosterone replacement therapy.

Since the late '90s, more and more studies were done demonstrating improved survival rates when various medical conditions are treated with TRT. Medical research has shown us that not only can men with low testosterone be safely treated with TRT but also they will have substantial improvements in quality of life and various other health parameters. We'll talk more about this later.

As more men with low testosterone were treated and physicians became more aware of testosterone deficiency symptoms, insurance companies started to take notice and began putting restrictions on testosterone prescriptions. Then two flawed studies were published, which were the first studies to contradict hundreds of prior studies that showed cardiovascular safety and cardiovascular protection in conjunction with TRT. The FDA acted with lightning speed and ruled that drug manufacturers had to change the labeling of testosterone to identify which types of testosterone deficiencies the FDA felt were appropriate to treat; labels also had to state that there was a possibility that testosterone replacement therapy could increase cardiovascular events.

When those studies were proven to be incorrect, the FDA reissued its statement, acknowledging that there was no scientific basis behind its ruling . . . but, nonetheless, the agency let its ruling stand. The FDA forced drug manufacturers to discontinue marketing testosterone as a treatment for low testosterone, specifically stating that men with symptoms of low testosterone should not be treated for low testosterone because that's simply part of getting older.

As of the publication date of this book, a series of lawsuits against manufacturers for their testosterone products are in the courts. These lawsuits have stoked fear among physicians who would otherwise want to prescribe testosterone to their patients. Even if doctors know every detail of the scientific literature, they don't want to get sued by their patients when the doctors are acting in their patients' best interests.

In the next chapters, I'm going to recap this scientific literature. We will talk about things that are definitive, things that are probable, and things that are questionable. That way, you can make your own decision when it comes to your health care, particularly testosterone replacement therapy.

The doctor-patient relationship has evolved over the years— nowadays, men are more interested in their health and how they can prevent disease. In short, men want to know what their choices are. This book gives you an opportunity to be fully informed.

# 2

## What the FDA Doesn't Want You to Know about Testosterone Replacement

The FDA approved testosterone in 1953. The first approved use of testosterone for women was an intramuscular injection, or shot. Testosterone pellets were approved in 1972; testosterone gels were approved in the '90s. Since that time there have been other approved delivery forms, including intranasal gels and one you put on the inside of your cheek to dissolve. There are no oral testosterone formulations currently available in the United States because the gastrointestinal absorption process converts it into a substance that is toxic to the liver.

1.  *The FDA wants you to think that only men with damage to their testicles or brain or those who have an extra female chromosome should consider testosterone replacement.*

The FDA wants to restrict testosterone replacement to men who have such low testosterone levels that "regression of secondary sexual characteristics, impaired sexual function, impaired sense of well-being, muscle wasting and decreased strength, and reduced bone mineral density" occur. The FDA requires that men with these rare conditions demonstrate muscle wasting and shrinking genitals before they can be treated with testosterone.

The FDA considers testosterone replacement to be restricted only to men who have damage to the testicles (from injury or infection) or injuries to the pituitary gland of the brain (caused by "tumors, trauma, or radiation"). In addition to damage to the testicles or brain, the FDA allows men with an extra female chromosome to utilize testosterone if their testosterone is severely low and markedly damaging their health. Even in these very restricted cases, the FDA wants the blood tests for these men to be double-checked!

*Bottom line: the FDA doesn't want men to know that if they have low testosterone, regardless of cause, they can benefit from TRT.*

## Science prevails

In this book, I review numerous studies that show the benefits of TRT when provided to men with low testosterone who have normally declining testosterone levels. None of these studies were singling out men who meet the FDA's criteria of having damage to the testicles or the brain.

If we were to follow the FDA rules and *not* prescribe testosterone for men with low testosterone secondary to aging (remember that the rules are for manufacturers but are intended to influence doctors and consumers), the largest studies done on this condition demonstrate that men with low testosterone would have double the risk of cardiovascular disease and premature death!

2.   *The FDA does not want men who are overweight or obese to know that testosterone replacement leads to substantial and sustained weight loss.*

Studies consistently show that even in severely obese men, TRT leads to profound, persistent weight loss and reduced waist circumference.

Testosterone replacement in men who are overweight or obese and who have symptoms of low testosterone is far more effective than any drug the FDA has approved for weight loss to date. In the studies we'll be talking about, testosterone has been shown to be universally effective in the treatment of obesity.

3.   *The FDA does not want men with diabetes to know that testosterone replacement improves and can even resolve type 2 diabetes.*

Clinical studies show that testosterone not only significantly improves markers of diabetes (such as elevated blood sugar, elevated hemoglobin A1c, and increased weight) but also, in some cases, reverses diabetes. None of the drugs the FDA has approved for diabetes actually reverse diabetes.

The FDA has acknowledged that testosterone replacement in men with diabetes leads to better control of blood sugar. Still, the FDA feels this is not an adequate rationale for using TRT because there are other drugs available to treat diabetes; also, the FDA says, many men with diabetes do not have brain or testicle damage.

4.   *The FDA does not want men with cardiovascular disease to know that testosterone replacement reduces strokes and heart attacks!*

Studies have demonstrated that TRT markedly reduces a man's risk of a heart attack if he has low testosterone. That's because testosterone improves circulatory blood flow, thereby reducing the chance of heart attacks and strokes. Not surprisingly, men with low testosterone have a higher incidence of strokes and heart attacks. They are also more likely to be overweight, have diabetes, have high blood pressure, and have abnormal cardiac lipids, all of which are risk factors for recurrence of stroke or heart attack.

Testosterone replacement therapy has been shown to specifically improve each risk factor, including improved blood pressure, blood sugar levels, weight, cardiac lipids, and inflammation. TRT also improves survivability in men who are at risk for heart attacks. Yet despite all of this, the FDA has frightened doctors by labeling testosterone products as potentially increasing the risk of heart disease even though the FDA acknowledges there is no scientific rationale for its labeling.

5.    *The FDA does not want men with sexual dysfunction to know that testosterone replacement improves libido, sexual function, and erections.*

Studies show strong correlations between low testosterone, sexual function, and other metabolic diseases. These same studies also show that—along with improving sexual function—other metabolic diseases like metabolic syndrome and diabetes also improve when men with low testosterone are treated with TRT. Extensive long-term evidence demonstrates that TRT is beneficial to men's sexual health.

6.    *The FDA wants you to think that testosterone can cause liver damage.*

Studies show that the available forms of testosterone are perfectly harmless to the liver and that TRT can in fact lead to improvements in liver diseases such as fatty liver. Yes, testosterone actually *improves* the liver, but the FDA has labeled testosterone as potentially harming the liver even though the agency acknowledges that damage doesn't ever occur. Let me say this again: in the agency's own documentation, the FDA states that there is no approved form of TRT that is harmful to the liver, but despite that statement, the FDA requires labeling that says TRT may harm the liver. If this sounds confusing, that's because the situation is factually absurd. As stated earlier, oral TRT can lead to liver issues, but there are no oral forms of testosterone in the United States.

7.    *The FDA does not want you to take testosterone replacement for depression or mood disorders.*

Science shows that men experiencing depressive moods, various reduced senses of well-being, and/or a low sensation of vigor will experience an improved quality of life with TRT. We will talk about this more in a later chapter. It's true: the FDA doesn't want you to know that TRT has a positive effect on your mood.

8.  *The FDA wants you to think that testosterone can cause blood clots.*

A study following more than 70,000 men for a long period showed no relationship between testosterone replacement and blood clots. There's no evidence that the two are connected, yet the FDA requires labeling that misrepresents TRT as potentially causing blood clots. This is a myth.

9.  *The FDA wants you to think that testosterone replacement causes prostate problems.*

In numerous studies, when TRT is given, prostate function scores improve. There is absolutely no link between prostate cancer and testosterone replacement. Despite this, the FDA wants you to think that TRT causes prostate problems, and its ruling requires that testosterone be labeled with a warning that it does.

10. *The FDA wants you to think that testosterone can negatively affect your blood pressure.*

Numerous studies show consistent favorable effects on both diastolic and systolic blood pressure with long-term testosterone use. Still, the FDA doesn't want you to know that TRT can improve blood pressure in men with hypertension—TRT must be labeled as potentially negatively effecting blood pressure.

11. *The FDA wants you to think that testosterone replacement can negatively affect your cardiac lipids.*

Numerous studies show that testosterone replacement improves not only other features of cardiovascular risk but also (fairly consistently) the ratios of good and bad cholesterol. The FDA doesn't want men with high cholesterol and low testosterone to know that TRT reduces a man's risk of a heart attack and minimizes the risk factors for heart disease through the means of generally improving cholesterol.

12. *The FDA wants you to think that testosterone replacement therapy is related to congestive heart failure.*

Studies show that TRT improves features of congestive heart failure for men with the preexisting disease. Also, there is absolutely no evidence that testosterone can cause heart failure. Testosterone is clearly cardiovascular-protective—it reduces cardiovascular disease and mortality in men with low testosterone. The FDA doesn't want men with heart failure to know that TRT will likely benefit them.

13. *The FDA wants doctors and men to believe that only men with the most severely low testosterone levels will benefit from testosterone replacement therapy.*

The International Expert Consensus Resolutions state: "There is no testosterone concentration threshold that reliably distinguishes those who will respond to treatment from those who will not." Despite the FDA's definition of low testosterone, there is an abundance of evidence that shows that the arbitrary cutoff of 280–300 ng/dL does not correlate with symptoms of low testosterone in men. Yes, most men with testosterone levels that severely low will be symptomatic, but even men with levels in the 400 and 500 ng/dL range may show symptoms of low testosterone and can benefit from testosterone replacement therapy. The only way to reliably determine which men will benefit from testosterone replacement is to assess the response to testosterone therapy in men who are symptomatic. The FDA doesn't want you to know that your blood testosterone doesn't have to be severely low for you to benefit from TRT.

14. *The FDA wants doctors to avoid prescribing testosterone replacement therapy for older men.*

Science clearly shows us that older men are more susceptible to things such as diabetes, heart failure, heart attacks, and strokes. The FDA allows these conditions to be treated, so why not allow low testosterone secondary to aging to be treated? Studies specifically evaluating testosterone replacement in older men show substantially reduced rates of fatality and universal protection against heart attacks and strokes.

The International Expert Consensus Resolutions identify that there is no age at which testosterone replacement should be avoided. A study of more than 500 men ages 65–84 who were followed for six years resulted in the same beneficial improvements that younger men receive. We've already discussed that long-term testosterone replacement in older men substantially reduces the risk of all-cause mortality and also results in fewer heart attacks and strokes. All of this said, there is no reason to withhold testosterone for men as they age.

15. *The FDA wants the medical community to avoid treating symptomatic men with low testosterone if their testosterone is low because of aging.*

There is no evidence to support this position by the FDA. Virtually all the studies showing how TRT improves various medical conditions do not exclude men with age-related low testosterone. The American College of Endocrinology is at odds with the FDA on this position; in fact, there are no medical societies that agree with the FDA on this unusual age requirement. Not treating low testosterone related to age is as absurd as not treating hypertension, diabetes, or heart disease because they occur more often as men age. This attitude just makes no sense. The FDA cites research that shows that the majority of older men have low testosterone and then goes on to use this as a reason to recommend against treating low testosterone secondary to aging.

The FDA acknowledges that about 90% of men with testosterone deficiency who are symptomatic are not currently being treated with testosterone. That scenario aligns with the FDA's (very limiting) rules.

16. *The FDA identifies that the following conditions are associated with testosterone deficiency regardless of cause but feels these men should not be treated with testosterone replacement.*

The FDA acknowledges that the following conditions are associated with testosterone deficiency:

- Diabetes, obesity, insulin resistance, belly fat, and metabolic syndrome

- Abnormal cardiac lipids and cardiovascular disease

- Musculoskeletal disorders such as muscle wasting, bone thinning, and frailty

- Depression and mood disorders, Parkinson's disease, and Alzheimer's disease

- Sexual dysfunction

- Chronic pain and drug abuse

But even though the FDA has identified these concomitant conditions, it has ruled that testosterone replacement should not be used in these circumstances, unless there is an associated brain or testicle disease. The FDA acknowledges that these conditions increase as men age. However, men with lower testosterone levels also have a much higher incidence of these diseases. Still, though, the FDA maintains its position that if testosterone deficiency is age related, men should not be treated with testosterone replacement in spite of the fact that it leads to the disease states that they have identified.

The FDA feels as though inadequate scientific research has been done and that conditions like sexual dysfunction, depressed mood, decreased cognitive abilities, and fatigue are not meaningful conditions because the FDA has not "validated outcome instruments" to measure them. In other words, the FDA is acknowledging that TRT has been scientifically demonstrated to improve these conditions but that the very nature of this scientific process has not been validated by the FDA.

*17. The FDA wants you to think that there is inadequate research regarding testosterone replacement in men with low testosterone.*

The FDA states that there is an absence of large studies (the agency defines "large" as including more than 200 individuals) lasting over one year, a lack that makes it "difficult to interpret the long-term clinical impact of TRT (testosterone replacement therapy)." Yet, as we have discussed, there have been studies with hundreds, thousands, and tens of thousands of men that were conducted over the course of six, eight, and 12 years. By ignoring these studies and instead selecting only a small handful of (often flawed) studies to use for its decision making, the FDA doesn't want you to know that the science behind TRT is exhaustive and extensive.

*18. The FDA states that there is a lack of data consistently supporting testosterone replacement that measures lean mass and fat mass along with bone mineral density.*

Despite the FDA's statement about this lack of data, in the agency's own position papers, the FDA cites numerous examples of clinical evidence showing that low testosterone is specifically associated with losing lean mass, gaining weight and fat mass, and losing bone density. The FDA doesn't want you to know that TRT is healthy for your musculoskeletal system. What's even more interesting is that the FDA recommends that TRT be initiated only in a unique subset of men who have rare disorders and severely low testosterone levels; these men must also have muscle wasting and bone loss. In a nutshell, the FDA wants to require that your testosterone be so low that you have muscle wasting . . . and then suggests that TRT may not work, anyway.

## In summary

The FDA maintains a significant bias toward testosterone replacement. While the FDA does not regulate the practice of medicine, it does regulate drug manufacturers, and in this case, the FDA is requiring drugmakers to label their products with dangers that do not exist. The FDA has also forced drug manufacturers to stop doing public awareness campaigns that were making more men aware of their health options.

The decision of whether you can take testosterone replacement rests with you and your physician or healthcare provider. Testosterone is readily available in many forms. Low testosterone is easy to measure using a combination of clinical evaluations and laboratory values. And while an absolute blood testosterone level does not reliably predict who will benefit from testosterone replacement, it does aid in the clinical decision-making process and with monitoring a patient's response to the therapy.

# 3

# How to Diagnose
# Low Testosterone

One of the reasons millions of men with low testosterone are not being treated is because physicians misunderstand how to diagnose it. Let's review how a diagnosis is made. First, physicians are trained to take a patient's history and do a physical exam. Gathering a patient's history is done by collecting data the physician gets from asking open-ended questions about the patient's concerns along with other parts of the patient's history, such as whether the patient has had surgeries (and what kind), what kind of medical issues the patient may have had/still has, how much alcohol the patient drinks, and whether the patient has ever smoked. Then the physician asks more specific questions relating to the patient's symptoms before doing a physical exam.

This leads to a provisional diagnosis. In some cases—like a common cold—your physician will tell you to take something over the counter, and no additional data are needed. By the time a patient history is taken and a physical exam has been done, the physician has obtained 90% of the data he or she needs. This is why the diagnosis is provisional rather than definitive.

Then come additional data, such as blood tests, X-rays, or an EKG as needed. In some cases, if the physician suspects you may be having a heart attack, he or she will immediately order an EKG. If the EKG is positive, it will confirm the provisional diagnosis of a heart attack. If, however, the EKG is negative and your physician still suspects you may be having a heart attack, he or she will then treat you as if you may indeed be having a heart attack—you may be given aspirin, oxygen, and other supportive measures while additional workup is performed.

Here's what your physician will *not* tell you: "During your history and physical exam, I noticed you were a little pale, sweating, and clutching your chest, so I was convinced you're having a heart attack, but seeing as your EKG came back normal, I'm sending you home. Good luck!" This will never happen.

. . . Yet this *will* happen when it comes to a man who is experiencing symptoms of low testosterone. As a medical community, we have become way too reliant on an absolute blood test as being the only sign of low testosterone. Remember that there are numerous factors that affect testosterone in the body, and there are proteins of hormones that interact with testosterone. The total testosterone level is not a reliable predictor of who will respond to testosterone replacement therapy and who won't. Certainly, absolute testosterone should be checked (along with other things), but the testosterone level alone should not be the sole criteria for determining whether a man has low testosterone syndrome and whether he will respond to testosterone replacement therapy.

The government, certain health insurance companies, and certain medical groups continue to assert their claim that testosterone blood level—that is, total testosterone level—is the sole determinant that should be used to justify testosterone replacement therapy. As we've discussed, the FDA has added to this the requirement that men have a genetic disorder or injury to the brain or testicles in order to qualify for testosterone replacement therapy.

However, more forward-thinking, modern, and updated guidelines tell us that there is no threshold that reliably predicts which men will benefit from testosterone replacement therapy and which men will not. It is reasonable to check the level, yes, but it is *not* reasonable to use the level as a sole factor for determining whether testosterone replacement therapy should be prescribed to men with symptoms of low testosterone.

Remember that testosterone deficiency is officially (and mistakenly) defined as happening when a man's testosterone level drops below a reference range on a laboratory test. When it comes to some aspects of medicine, this concept holds true—for instance, when your doctor does a CBC (complete blood count), there are reference ranges that identify the possibility of infection or anemia. However, when it comes to testosterone levels, a strict cutoff level is not a good indicator.

As physicians, we are trained to make a diagnosis based on three factors. Number one is taking the patient's history (as described above) and a complete set of symptoms. Symptoms might be weakness or fatigue, for example, or a sore foot, a sore throat, a headache, or a bump on the skin. Doctors are trained to ask their patients questions in an open-ended fashion to figure out what the patient's symptoms are and what the next step should be. Taking a patient's history is generally considered to result in collecting about 70% of the data needed to make a medical decision and a diagnosis. Knowing the patient's history is essential for determining what is likely the cause of the patient's concern.

The next step is the physical exam. The exam may be extensive and whole body, such as doing an annual physical exam, or it may be limited to something like looking at a spot on the patient's skin. The physical exam is generally considered to constitute approximately 20% of the data gathering and medical decision making.

The last step is getting the test data. Test data can be laboratory values, X-rays, various imaging studies, and different test results.

This makes up about 10% of the data gathering and medical decision making. This 70-20-10 ratio might sound unbalanced, but that's how a diagnosis is determined.

You may think, "But if the X-ray tells a doctor whether my arm is broken, that's definitive—it's definitely broken, or it's definitely not broken." This is true. Yes, an X-ray will show that the arm is broken or not broken. But how does the doctor determine that an X-ray should be done to check for a broken arm as opposed to having an ultrasound done to check for a blockage in the artery? The decision as to which testing method to use is based on taking the patient history and doing the physical examination.

So how do we determine whether a man has low testosterone? How do we know who will respond favorably to testosterone replacement? Even though insurance companies use very strict, rigid guidelines to determine whether they will pay for the testosterone replacement, there is no factual basis for this. The normal "reference range" for testosterone level is between 250 and 1,100 ng/dL. It's called the reference range because that's more or less the range of values for 95% of men who are generally healthy. It's just a statistic. And the "normal" range was established early in the past century.

A study was published in *Aging Male* in 2015 detailing the initial results of the UK Androgen Study (UKAS). Androgen is the category of hormones in which testosterone is the dominant player—it's more or less considered to be synonymous. The study went on for 25 years and followed thousands of men, which was an enormous undertaking.

When the UKAS started back in 1990, diagnostic guidelines for testosterone replacement had not yet even been established. In other words, the researchers began studying the response that men have to testosterone treatment before the FDA or any other agencies had redefined how a "low testosterone" diagnosis should be defined, back before there were consistent guidelines established by the government (and others) as to who should be treated with testosterone and who should not be.

In the UKAS, researchers based symptoms of low testosterone on the Aging Male Symptoms (AMS) scores. This scoring system—along with another one called the ADAM questionnaire—was considered to be a reliable indicator of low testosterone.

Men with symptoms of low testosterone based on their AMS scores were given testosterone replacement. In this study, the most consistent symptoms of testosterone deficiency were loss of libido, decreased energy, erectile dysfunction, loss of morning erections, night sweats, joint pains, depression, irritability, and impaired memory. The men had been experiencing symptoms for three to five years prior to seeking treatment for low testosterone. Their testosterone levels ranged from about 150 to 1,100 ng/dL, with an average of just over 450 ng/dL.

Remember that when symptoms are present, absolute testosterone levels cannot be directly attributed to low testosterone. In fact, only 8% of men who have significant quality-of-life symptoms that are typically thought of as being related to low testosterone have a testosterone level lower than 250 ng/dL; therefore, these men don't meet the official threshold for "low testosterone." That means that 92% of men with low testosterone symptoms have levels that are considered "normal" by the FDA and others.

Back to the study involving thousands of men who were followed for 25 years. Forty-two percent of those men had a testosterone level greater than 460 ng/dL. The majority of nonenlightened physicians would consider this level to be perfectly normal and would dismiss their patients' symptoms as being insignificant. For the most part, in fact, the men in this study were all in the normal reference range (which dips as low as 250 ng/dL).

The men in the study with symptoms of low testosterone as defined by abnormal scores on the AMS questionnaire were offered testosterone replacement. (Some men in the study who had symptoms that were not likely related to low testosterone were not treated.) The researchers found that regardless of their initial testosterone level, the men's AMS scores consistently dropped over the first year, meaning their symptoms went away . . . and their symptoms continued to be nonexistent for over a decade. This offers some of the best real-world proof that while we should check men's total testosterone levels, those levels are largely meaningless. (Remember that the men's testosterone levels in this study ranged from about 150 to 1,100 ng/dL, with an average of just over 450 ng/dL.)

Physicians who understand that the total testosterone level is close to meaningless frequently do not check testosterone levels. In fact, several studies show that about 70% of men who are put on testosterone replacement therapy never had their testosterone level checked in the first place. There's a very good reason for this. Because there's so much confusion about what the testosterone level means, physicians who have expertise in testosterone replacement often choose *not* to check the total testosterone level to avoid confusing their patients *and* their peers.

The flipside is most physicians who are not particularly versed in testosterone replacement therapy do check the total testosterone and base their entire decision of whether to treat solely on the total testosterone level. I don't think either method—completely ignoring testosterone levels or relying solely on them—is the correct way to monitor and manage symptoms of low testosterone. Although I completely understand why some physicians don't check it in the first place, in my practice, a man's initial testosterone level sets a baseline. One of the things we've been looking at in these various studies is that testosterone replacement is not just about replacing the testosterone to a certain level but also about replacing the testosterone to a more ideal level. For example, if a man starts out at 250 ng/dL, he will probably feel pretty good at around 600, but if a man starts out at 400 ng/dL, he probably won't feel good until about 800. That's why I use testosterone levels as a guide and not as a rule.

The International Expert Consensus Resolutions that we discussed earlier (published in the *Mayo Clinic Proceedings* in July of 2016) summarizes the panel's unanimous consensus on this issue. In resolution 5, the panel members emphasize: "There is no testosterone concentration threshold that reliably distinguishes those who will respond to treatment from those who will not." They specify several variables that make the testosterone threshold level an unreliable indicator. Blood work is *not* a good way to test for low testosterone.

## In summary

To be clear, I am not suggesting that testosterone levels not be checked—we check them when therapy is initiated and see how they coordinate with patient symptoms. As we previously discussed within the context of the 25-year-long UKAS, men can be symptomatic at any point along the normal reference range; likewise, the reference range does not determine which men will respond to testosterone replacement therapy. That said, there is somewhat of a correlation. If a man has symptoms of low testosterone and his testosterone level falls into the midrange or above, I am more likely to consider other causes and may do additional diagnostic tests—for example, a man with fatigue and low energy who has a testosterone level of 1,100 ng/dL should be investigated for things such as anemia. When men have symptoms of low testosterone, we do a workup that that includes more than just measuring testosterone levels.

The ultimate test to determine whether a man who has symptoms related to low testosterone will respond to testosterone therapy is very simple: replace testosterone and remeasure his symptoms. If the symptoms diminish, it's clear that particular individual will respond to testosterone replacement therapy. In reality, if a man has symptoms of low testosterone without any other identified cause, he will almost always respond to testosterone replacement therapy. Michelle, who has worked in our hormone replacement division for about 10 years, says, "I wish young healthy men would have their testosterone checked and kept in their records so we would know at what level of replacement they will feel their best."

# 4

# Testosterone and Weight Gain

In this chapter, we will talk about the relationship of low testosterone to weight gain and obesity. The links are extraordinary. We'll cover this relationship from a cause-and-effect standpoint—in other words, does low testosterone cause obesity, or does obesity cause low testosterone? More on this later, but it's a circle. The two are interconnected; either one can cause the other. We will also talk about metabolic syndrome, which is a prediabetic condition.

The chapter on diabetes will review the strong correlation between low testosterone and diabetes and vice versa. But before a man gets type 2 (or "adult-onset") diabetes, he will generally gain weight, particularly in the midsection—that is, he'll develop belly fat, or what's called a "beer belly." He will start developing a syndrome called metabolic syndrome, which is a condition where we see elevated blood pressure, a rise in cardiac lipids such as cholesterol and triglycerides, and a rise in blood sugar as it starts nudging up toward but isn't quite yet at diabetic levels. Metabolic syndrome occurs before diabetes happens.

There are various causes of weight gain and obesity. In this chapter, we will focus on the relationship of low testosterone to weight gain and obesity. Note that I'm not trying to rule out other causes of obesity such as decreasing physical activity (which seems to be inevitable as society evolves) and our changing appetites and food choices. If you have gained weight, you obviously have to address these aspects of your lifestyle. However, in this chapter, we'll talk about the very strong contribution low testosterone makes to weight gain and the relatively extraordinary and generally misunderstood effect that testosterone replacement has on reducing obesity.

## First, let's go over some terms we will discuss in this section:

*BMI*

BMI stands for body mass index. This is a common measurement of relative leanness versus obesity. The formula used to determine BMI considers only your height versus your weight, however, so someone who is very muscular may have a BMI value that qualifies as "overweight," but in reality, that person may have very little fat. Likewise, someone with low muscle mass may have a big potbelly but a normal BMI. All of that said, *generally speaking*, BMI is a way to monitor weight loss/gain trends and—though an imperfect measurement—it is fairly helpful. You can Google "BMI calculator" and check yours.

- Normal BMI is considered 18.5 to <25.

- Overweight is 25 to <30.

- Obese is 30 and over.

- Class I obesity is 30 to <35.

- Class II obesity is 35 to <40.

- Class III obesity is 40 and over.

*Metabolic syndrome*

Metabolic syndrome is sometimes also called syndrome X. The condition refers to having a set of risk factors that raise your risk for diabetes, heart disease, and stroke. The best way to think of metabolic syndrome is as "prediabetes," because if changes are not made, diabetes follows. These risk factors are generally defined as central obesity (a.k.a. belly fat), high blood pressure (whether treated or not), abnormal cardiac lipid blood tests (such as high triglycerides and low HDL), and a fasting blood sugar level that's elevated but not yet at a diabetic level.

*Type 2 diabetes*

This is also known as "adult-onset" diabetes and "non-insulin-dependent" diabetes. Type 1 diabetes is a related but different condition. We are not going to talk much about that type, but briefly, type 1 diabetes is a disorder of blood sugar regulation that is unrelated to weight gain. Generally, type 1 begins in childhood—it used to be called "juvenile diabetes"—but sometimes adults can develop type 1 diabetes as well.

Type 2 diabetes is a condition where repeated exposure to sugar in the blood causes the body to make more and more insulin to drive the sugar into the cells and out of the blood. If the sugar were to remain in the bloodstream, the blood would get "sticky" and would lead to an increased risk of heart attack, stroke, and permanent capillary damage (among other conditions). Eventually, though, the cells resist the insulin, which means that the blood sugar does remain dangerously elevated.

## Trends in weight gain

From 1960 to 1980, the average American maintained a pretty stable body weight. Since then, however, weight gain and obesity rates have

gone up dramatically. You can get more details about this trend at www.
niddk.nih.gov/health-information/health-statistics/Pages/overweight-
obesity-statistics.aspx.

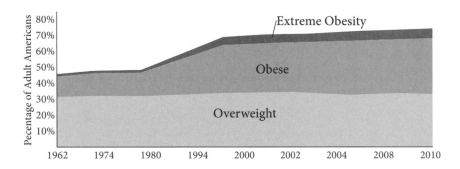

As you can see from the above chart, from 1980 to the late '90s,
rates of obesity and extreme obesity skyrocketed and have held there
ever since. In 1962, about 32% of adult Americans were overweight; in
2010, about 34% were. Not much change here. But while only 13% of
American adults were obese in 1962, starting in the late '90s (and continuing until now), that rate has become about 36%, meaning that the
number of obese Americans has essentially tripled. During that same
time frame, extreme obesity has gone from 1% to 5%, which is a 500%
increase.

Extensive research is ongoing to determine what is behind this
spike in average weight gain. Fortunately, the trend is slowing down. We
are still getting heavier as a society, yes, but the rate of weight gain has
slowed.

Numerous studies have linked low testosterone to weight gain, obesity, prediabetes, metabolic syndrome, and diabetes.

## Low testosterone and obesity

Low testosterone and obesity are not just American problems. Global
trends toward obesity—as well as the link between obesity and low

testosterone in men—are being investigated around the world. For more than 30 years, we have known that there is a link between low testosterone and obesity. We will discuss the relationship first, then talk about the clinical results of TRT's effects on obesity, metabolic syndrome, and diabetes.

A study published in the *Journal of Diabetes Research* in 2016 looked at middle-aged Polish males with metabolic syndrome and obesity and compared them to controls, meaning men with no obesity or prediabetes. The researchers found a more or less linear relationship between declining testosterone levels and increasing abdominal girth and the development of prediabetes.

Another study done in Nigeria, published in 2104 in *International Endocrinology and Metabolism,* looked at men with metabolic syndrome and compared them to men without metabolic syndrome. The study found that not only did the majority of men with metabolic syndrome have low testosterone but also the men with belly fat were about eight times more likely to have low testosterone than men without belly fat. Researchers also found that men with obesity, metabolic syndrome, hypertension, or elevated cardiac lipids (cholesterol, etc.) are more likely to have low testosterone than men without any of these conditions.

In the European Male Aging Study (EMAS), the largest study of aging in the world—it covered men in eight different countries—the researchers found that overweight men are about three times more likely to have low testosterone, while obese men are about nine times more likely to have low testosterone. This is pretty much in line with other population studies—additional studies done in Chinese and Australian men show the same pattern.

In the Massachusetts Male Aging Study (MMAS), researchers noted that as men become obese, the loss of their testosterone is equivalent to suddenly aging 10 years in terms of what their predicted natural reduction of testosterone levels would be.

We have definite evidence that men who are obese or prediabetic are more likely to have low testosterone. Does obesity cause low testosterone? Or does low testosterone cause obesity?

This may be hard to definitively answer, because they go hand in hand. This bidirectional link is explained in the article titled "Metabolic Syndrome and Hypogonadism—Two Peas in a Pod," published in *Swiss Medical Weekly*.

Various studies about this bidirectional link have been done in healthy young men. The results? When researchers gave the young men chemical testosterone blockers, their fat mass increased immediately, but when the testosterone blockers were removed, the men saw a return to their normal weight.

Other studies looking at baseline testosterone levels that followed men for seven-plus years have seen a trend of eventual abdominal fat mass increasing in men with lower testosterone levels even when they are lean at the beginning of the study. So, even if you are lean right now, if your testosterone is low and you have symptoms of low testosterone, you will likely gain weight if you do not treat your low testosterone with testosterone replacement.

In short, low testosterone and obesity seem to have a bidirectional effect: low testosterone predicts obesity, and obesity predicts low testosterone. There are likely other variables at play as well. Low testosterone is associated with reduced muscle mass, for example, which can lead to reduced exercise performance and less calorie burn, hence more fat. Low testosterone also leads to quicker fatigue when doing aerobic exercise. Low testosterone leads to lowered sexuality, too, which in itself is fat burning and may also be a motivator for taking better care of oneself.

## Does weight loss have a beneficial effect on testosterone level?

The answer is easy: yes, weight loss does improve testosterone. In fact, several studies have shown that significant weight loss can improve testosterone levels. And there seems to be a linear correlation, too—modest weight loss leads to modest improvements in testosterone, whereas significant weight loss leads to significant improvements in testosterone levels. If you have low testosterone and are overweight, making lifestyle changes will be part of the plan.

## Does testosterone replacement lead to weight loss?

*Testosterone replacement in overweight men*

A study was published in *Clinical Obesity* in 2013 that detailed a long-term observation of obese men receiving testosterone therapy (they were followed for up to five years). The men were 32–80 years old, with the majority being middle-aged or older. In this study of men receiving testosterone replacement, only 4% of the men were normal weight. Ninety-six percent of the men in this study were overweight or obese. The cutoff testosterone level for men receiving TRT in this study was below 300 ng/dL. If the researchers had used a cutoff of, say, 400 ng/dL, there would likely have been more normal-weight men taking part in the study. (Remember that the lower his testosterone is, the more likely the man is to be overweight.)

To recap, 4% were a normal weight, 34% were overweight, and 62% were obese.

In this six-year study, the researchers saw a consistent year-over-year reduction in waist circumference among the men receiving TRT. In fact, an astonishing 97.5% of the men saw a reduction in their waistline, and about half of the men lost a full five inches off their waistline.

And they lost weight in the same fashion year over year. In fact, the longer they were on testosterone replacement, the more weight they lost—every year, more than 50% of the men lost more than 20 pounds, and a significant number lost more than 40 pounds. Only 4% of the men gained weight in this six-year-long study. Pretty astonishing.

## Testosterone Level

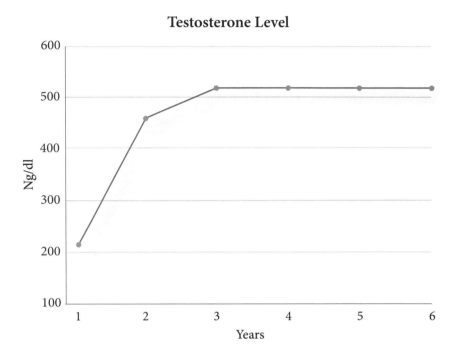

## Waist Circumference

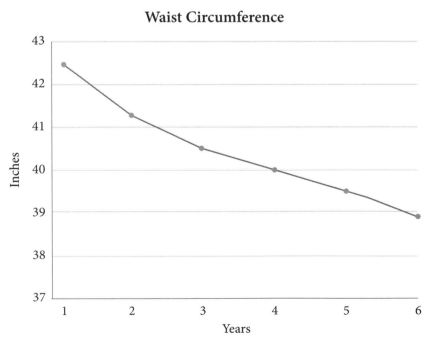

Another study published in *International Journal of Obesity* in 2016 analyzed these rates of weight loss. It was titled "Effects of Long-term Treatment with Testosterone on Weight and Waist Size in 411 Hypogonadal Men with Obesity Class I-III: Observational Data from Two Registry Studies." In this paper, the researchers identified men who are both low testosterone and obese as being defined by having a BMI of 30 and over. (As previously mentioned, class I is BMI 30 to <35, class II is 35 to <40, and class III is 40 and over.) The men were followed for eight years. The results follow:

> **Class I obesity.** These men lost an average of 38 pounds during the eight years and did so steadily, losing weight gradually every year. The majority lost more than 10% of their body weight. Of the more than 200 patients in this group, only three of them gained weight in the eight years.

> **Class II obesity.** In this group, the results were more stunning: they lost an average of 56 pounds during the eight years! They also lost several inches off their waistline. In this group—which consisted of 147 men followed for eight years—not one of the men who received testosterone replacement had weight gain. Not a single one.

> **Class III obesity.** These men lost an average of 67 pounds during the eight years of taking testosterone replacement. Again, this group lost several inches off their waists, and not one man gained weight over the eight-year study period.

The men in these groups didn't just lose weight in this study: their cholesterol also improved, their fasting blood sugar improved, their triglycerides improved, and their blood pressure improved.

All in all, of the 411 men, only three gained weight in eight years. On average, the men lost 20% of their body weight. To put this in perspective—and to contradict the stance of the FDA ruling that testosterone replacement should be used only for men with very low testosterone caused by organ damage or an extra female chromosome—let's look at the how the weight-loss drugs the FDA has supported stack up.

## Drugs versus evidence-based medicine

Note that there are no FDA-approved drugs that have testosterone replacement's track record for weight loss, yet the FDA has ruled that men such as the 411 men in that study should not be offered testosterone replacement. The agency has, however, approved drugs that do not have the long-standing safety track record that testosterone has. Orilstat (Xenical), Lorcaserin (Belviq), and Phentermine/Topiramate (Qsymia) are a few of these products.

Orilstat works by blocking an enzyme and thereby blocking absorption of fats. This leads to modest weight loss over time—about 50% of people taking this drug for a year will see a 5% loss of body fat. Orilstat comes with the unpleasant side effect of "anal leakage" and greasy, runny stools.

Belviq was approved in 2012 by the FDA. This potentially habit-forming hallucinogenic drug helps about 40% of people lose 5% of their body weight and has substantial side effects.

Qsymia was approved in 2012 by the FDA. This product is a combination of a stimulant and an anticonvulsant, both of which have the side effect of appetite suppression. This drug combo has the potential to lead to 5% weight loss in 70% of the people who take it. At the highest dose—which comes with the most side effects—Qsymia can result in a weight loss of up to 9%.

Still, despite the side effects that come with FDA-approved weight loss drugs, the FDA has warned manufacturers that testosterone should not be used in obese men with symptoms (and laboratory evidence) of low testosterone. Yet in obese men with low testosterone, not only does testosterone replacement yield meaningful weight loss in one year but also it does so year after year and without major adverse effects. This cannot be said for any of the weight-loss drugs approved by the FDA. Not one of the agency's approved drugs comes close to those results.

## Back to TRT and weight loss

The issues and outcomes we are talking about here are what we call "evidence-based medicine." As physicians, we are supposed to keep up on current studies and adopt a medical practice that responds to the latest valid evidence. I am going to lay out numerous studies that support this contention; we will also talk about any evidence to the contrary. However, in terms of weight loss and testosterone, there isn't any contrary evidence—all the studies I reviewed (which are basically all of the abstracts and papers on the subject in peer-reviewed journals) come to the same conclusions. And the literature is tremendous in scope. I am going to present the studies that are particularly clear and relatable.

Another study published in 2011 was done at an institute in Las Vegas that specializes in treating men with low testosterone as well as men who are borderline low (versus the traditional practice of just treating men with very low testosterone). This study addressed weight loss in men who are not obese but rather just overweight; these men have only mildly low testosterone and want to lose a little belly fat.

The study included men with a BMI over 26 (so just barely overweight) who had minor symptoms of low testosterone. These symptoms can include the following: abdominal fat, lack of energy, fatigue, loss of libido, and/or erectile dysfunction. Again, these men had testosterone levels that weren't at the lowest level, but they were at a suboptimal level. The men were given dietary and exercise advice as well as testosterone replacement.

In this study, the men's average testosterone level was 437 ng/dL, which is considered normal by the FDA but which is well below the levels at which men typically feel normal. Each man's testosterone was replaced until it reached a level of about 900 ng/dL, which is a normal level in a man who has no symptoms of low testosterone.

At the one-year mark, they had lost about 10 pounds on average, with a total range of 1–47 pounds. (The group included 50 patients.) At the follow-up mark, all but two men had lost some body fat. There are no FDA-approved drugs that are this successful and safe for weight loss.

Another study, this one titled "Long-term Testosterone Therapy in Hypogonadal Men Ameliorates Elements of Metabolic Syndrome: An Observational, Long-term Registry Study," followed hundreds of men with an average age of 58 and with at least mild symptoms of low testosterone. These men also had laboratory evidence of low testosterone. The researchers used below 350 ng/dL (or 12 nmol/L) as the definition of "low."

In this study, the men also had metabolic syndrome. Basically, at the beginning of the study, they were a bit overweight with a beer belly, they had high blood pressure, and they had abnormal cardiac lipids such as high cholesterol.

During this five-year study, the men's testosterone levels went from an average of 286 ng/dL (9.9 nmol/L) to 519 ng/dL (18 nmol/L). They lost an average of three inches off their waist and about 34 pounds. Their cholesterol went from an average of 282 to 188 mg/dL (i.e., from high to normal). Their LDL, or "bad" cholesterol, went from an average of 164 to 110 mg/dL (again, from high to normal). These effects gradually accumulated over two years and remained stable for five years . . . and no lipid-lowering drugs were used! Their HDL, or "good" cholesterol, rose, which is beneficial, and their triglycerides fell by 32%. Their blood pressure changed from an average of 154/93 to 138/80 mmHg. In short, the TRT normalized their blood pressure. Both blood sugar and insulin levels improved as well. Markers of inflammation and liver fatigue also improved.

## Consistent evidence shows the benefit of testosterone replacement

The evidence consistently demonstrates that men with low testosterone respond to testosterone therapy with improvements: they see decreased body weight, increased fat loss, increased lean muscle mass, and improved metabolic parameters.

A Brazilian study evaluated short-term testosterone replacement—the men in this study with low testosterone received only six months of TRT. The goal was to study the effects of TRT on men who have had low testosterone for a long time, so the researchers chose men with an average age of 70 (it's commonly acknowledged that 50%–80% of men have low testosterone at this age). What they found is that replacing testosterone still improves waist circumference, body fat, and libido, even when given to men only short-term and when given to men who are 70 years old on average *and* who have likely had low testosterone for years. (For additional fascinating studies, please download the professional version.)

## In summary

These studies have shown us clear links between low testosterone and weight gain and testosterone and obesity. Weight gain and obesity are enormous health concerns in America and elsewhere. These factors are associated with the development of other diseases such as diabetes, cardiovascular disease (including myocardial infarction and stroke), high blood pressure, arthritis, and even cancer. Avoiding weight gain and obesity is paramount for all men. For men who *have* gained weight or gotten obese and who have low testosterone, testosterone replacement should be considered. There is no reason that a man who is gaining weight and has symptoms of low testosterone should avoid testosterone replacement. As we'll talk about in the following chapters, when testosterone is replaced to therapeutic levels, there are no adverse events. Also, it can be taken indefinitely. There are no other drugs we can say this about.

Weight-loss drugs are a big market—they are advertised every-where. Despite this, they really don't do much, and they carry significant side effects and inherent risks. In contrast, testosterone replacement consistently leads to weight loss in men who are overweight or obese and have low testosterone.

But know that we—you as a patient and I as the physician—are not regulated by the FDA. I have an ethical duty to do what's in the best interest of my patients and to evaluate the best evidence-based med-icine. This is how my staff and I practice medicine: I discuss the full situation with my patients so that we can determine what is their best option for their optimal health. I do not blindly follow the FDA's rules. I'm certainly aware of them and take them into consideration, but the FDA is not infallible, and, in this case, I do not prescribe the drugs the FDA has approved for weight loss. For men with low testosterone and weight problems, I prescribe testosterone replacement as the best, safest, and most effective option.

# 5

# Testosterone and Type 2 Diabetes

In the last chapter, we talked about low testosterone and its relationship to metabolic syndrome, which can also be considered prediabetes. We also reviewed the relationship between low testosterone and weight gain, which precedes metabolic syndrome.

Weight gain, particularly when it is belly fat, is strongly associated with the eventual development of metabolic syndrome. In men, metabolic syndrome is a condition where the waist is bigger than the hips. This happens in conjunction with elevated cholesterol or abnormal cardiac lipids, elevated blood pressure, and elevated blood sugar (that is, blood sugar that's not elevated—yet—to quite the diabetic level). Once metabolic syndrome occurs, the natural progression is a march toward type 2 diabetes (also known as adult-onset diabetes or non-insulin-dependent diabetes).

After diabetes comes cardiovascular disease (heart attacks, strokes, and heart failure), increased risk of cancer, and other causes of premature death. Clearly, there is a link to diet and exercise in the development of type 2 diabetes. There is also a clear association between low testosterone and diabetes. Even though other issues are critical to managing diabetes—for example, the importance of weight maintenance through lifestyle changes, including healthy eating and exercise—we're going to talk specifically about how low testosterone is related to diabetes.

Type 2 diabetes is very concerning to me on a personal level. Throughout my life, I've been extremely dedicated to watching my diet and exercising because of my fear of developing weight gain or diabetes. Both my mother and my father had diabetes at a relatively young age, and both my brother and sister developed type 2 diabetes at a young age as well. There wasn't much encouragement to exercise—if any—in my family. But despite my family background, I never developed type 2 diabetes, and I have remained very lean my entire life. I'm very consistent in terms of regularly exercising as well as being faithful to an antidiabetes diet.

## Which came first, low testosterone or diabetes?

We're going to focus on evidence of the connections between testosterone replacement and diabetes modification as well as whether diabetes is caused by low testosterone or whether low testosterone causes diabetes. In reality, they are probably bidirectional—there is evidence that low testosterone definitely leads to a diabetic environment. However, that particular issue has not been studied a lot. What has been extensively researched is the use of testosterone replacement therapy to treat diabetes.

## Clinical studies

A study published in *Clinical Endocrinology Oxford* in February 2012 evaluated men's insulin glucose response when testosterone deprivation occurs. (Testosterone deprivation can occur by using a drug that suppresses the production of testosterone.)

In this study, the researchers split men up into two groups. All of the men in both group 1 and group 2 were given what is called a "gonadotropin-releasing hormone antagonist" drug, which completely blocks the production of testosterone. This makes a man's testosterone go from normal to almost zero. The men in group 1 were put on the testosterone blocker alone; the men in group 2 were put on a testosterone blocker but also given testosterone replacement at the same time. That way, the researchers could determine whether there was a relationship between testosterone and the development of diabetes.

Both group 1 and group 2 consisted of normal, healthy men aged 18–55 with no chronic medical conditions and on no medications. They also had no diabetes, normal blood sugar levels, and normal testosterone levels. These were perfectly healthy men who were not obese.

Group 1 men were given a daily placebo, while group 2 men were given daily testosterone replacement. So again, the researchers were studying two groups of men with artificially lowered testosterone, and only one of the groups was given testosterone replacement.

At baseline (i.e., the beginning of the study), all of the men had normal fasting blood sugar and insulin concentrations and had perfectly normal insulin sensitivity. That means a perfectly normal response to blood sugar.

Four weeks into the study, group 1 developed an abnormal insulin response. In this short time period, they became insulin resistant, meaning that their bodies had to make more insulin to manage their blood sugar. This is experimental evidence that low testosterone leads to insulin resistance. The men who had their testosterone totally blocked but who were simultaneously given testosterone replacement maintained normal blood sugar, normal insulin levels, and normal insulin sensitivity.

The results of the study provided scientific evidence that testosterone plays a direct role in the development of insulin resistance and thereby diabetes, a finding that calls into question the decades-long assumed or presumed assumption that diabetes is solely related to excess carbohydrate intake in genetically predisposed individuals. And this is not the only study demonstrating this link—numerous studies have been done in animals as well. However, the purpose of this particular study was to help answer the question of "What came first, the chicken or the egg (i.e., the low testosterone or the diabetes)?" The study showed that depriving men of testosterone rapidly led to insulin resistance and that depriving testosterone and then replacing it led to normal insulin sensitivity.

Now let's move on to studies showing real-life applications of testosterone replacement in men who actually have diabetes.

A study published in the *European Journal of Medical Research* in 2014 compared treating type 2 diabetes with lifestyle intervention and diabetic medications alone versus using those same treatments while also using testosterone. In this study, men ranged from overweight to severely obese, and it was a randomized placebo-controlled study, meaning that neither the men nor the examiners knew which group was getting testosterone replacement.

In this study, the researchers found that all of the men had low testosterone, which is expected in men with type 2 diabetes who are overweight or obese.

After six months, the researchers found that the intervention with lifestyle changes and antidiabetic medications improved blood sugar, hemoglobin A1c, cardiac lipids, and blood pressure. Great! But the men who had testosterone added to their regimen did much better.

A study published in *PLoS ONE* on June 25, 2016, was conducted in Denmark. It was a 19-year-long study concerning the relationship of weight loss in type 2 diabetic patients and ultimate mortality. In other words, does weight loss lead to a longer life for people with type 2 diabetes? There were 761 patients with type 2 diabetes in the study. They were treated with therapeutic and intentional weight loss through a physician-supervised program and were followed for an additional 13 years. The study found that all-cause mortality (meaning death rates as well as cardiovascular disease and other weight-related diseases) did not change in the group that lost weight. While I'm 100% in support of losing weight when you have diabetes, I'm reviewing this study with you so you understand that weight loss might not be the only factor that is critical when treating diabetes.

A study published in the *European Journal of Endocrinology* in 2013 was done in the United Kingdom. This study also evaluated mortality in men with type 2 diabetes. In this study, the researchers checked the men's testosterone level, and some men with low testosterone were treated with testosterone replacement and some were not. A physician also followed them for six years.

There were 560 men in the study, and as expected, many had low testosterone. Although they were all offered testosterone replacement therapy, only about 25% of them were compliant and took it. The researchers also tracked these men and their mortality rates. As expected, men with low testosterone had increased mortality compared to men with normal testosterone. However, the men with diabetes and low testosterone who took testosterone replacement had a significant reduction in mortality versus the men with diabetes and low testosterone who did not take testosterone replacement.

Another study done at about the same time followed older men treated at Northwestern Veterans Affairs Medical Centers. Of the roughly 1,000 men identified with low testosterone, about 400 of those men initiated testosterone replacement. In the subsequent five years, 20% of the untreated men died compared to 10% of the men treated with testosterone. When comparing the men who were treated with testosterone and who were more likely to survive versus the men who were not treated with testosterone and were more likely to die, it's notable that the groups had the same rates of diabetes and heart disease and were about the same age—so, again, once men are identified as having low testosterone, the most reliable predictor of their survival is whether their testosterone is replaced.

Another study done in Italy and published in the *European Academy of Anthology* in 2010 was a meta-analysis. This meta-analysis evaluated type 2 diabetes and testosterone deficiency as well as type 2 diabetes and testosterone replacement. Again, a meta-analysis is a type of study that looks at similar studies to aggregate the data and come to conclusions. An individual study can be criticized for having a small number of subjects (or a limited number of other variables), but compiling several studies and comparing similar events allows us to draw more meaningful conclusions.

This meta-analysis began with an assessment of 742 medical journal papers that had been written about the connections between testosterone and type 2 diabetes in men. The researchers found 37 studies that could be used for comparison. Pretty consistently, they found that low testosterone predicts diabetes and that diabetes is strongly associated with low testosterone. They also consistently found that testosterone replacement therapy improves blood sugar control, weight, HbA1c, fat mass, and triglycerides. In summary, the researchers found that no matter what type of testosterone replacement was used (short- or long-acting creams, injectables, or pellets) and regardless of the duration of the study and whether testosterone was restored to optimal levels or low "normal" levels, testosterone replacement improves diabetes. Three things are clear: men with low testosterone are more likely to develop diabetes, men with diabetes generally have low testosterone, and testosterone replacement therapy improves diabetic parameters.

## Testosterone treatment with a track record

We've reviewed some studies that used suboptimal forms of testosterone in men with testosterone deficiency and type 2 diabetes, and we've also reviewed some studies that evaluated men with low testosterone and diabetes who were undertreated with testosterone replacement. Now let's talk about some studies that treated men with long-acting testosterone. The following two studies evaluated men with type 2 diabetes and low testosterone who were treated with long-acting testosterone replacement, similar to the pellets we use in the United States. You'll see that in these studies, the testosterone level went from low to normal—not from low to *low normal* but from low to *actually* normal.

A study published in 2014 in the *International Journal of Endocrinology* evaluated 156 men with type 2 diabetes who had an average testosterone level of about 200 ng/dL, which is very low. They were given long-acting testosterone replacement every few months. Their testosterone level was brought to over 450 ng/dL, which is not ideal but is closer to ideal than what we saw in the other studies we reviewed. The men in this study were an average age of 61, and they were all obese with a waist circumference of more than 37 inches. Their age and waist circumference were not required criteria for the study, but it's not surprising that they were of this age and size—that follows the typical pattern when it comes to men who have low testosterone and diabetes. Remember that both diabetes and low testosterone are associated with obesity and belly fat.

All of these men had abnormal cardiac lipids—in other words, they had elevated cholesterol. All but three had high blood pressure, which again we can expect because the onset of diabetes usually follows the onset of metabolic syndrome. Metabolic syndrome also includes hypertension, so (again) not surprisingly, about one-quarter of the men had some form of coronary artery disease. About 12% of them had previously had a heart attack. I hate to sound repetitive, but, again, this is not unexpected because heart attacks and coronary disease are the next natural stepping-stone from diabetes.

The men were followed for up to six years. The average waist circumference at the start of the study was 45 inches. (Every man's waist was at least 37 inches, so some men had a waist much bigger than 45 inches.) The results? The men lost an average of five inches off their waistline. They also lost an average of 39 pounds. Again, this is over six years, so each year there was an incremental loss, with their weight loss starting in year one. Additionally, their blood sugar fasting went from more than 140 (which is well within the diabetic range) to 110 (which is not in the diabetic range). Remember, this happened over six years, with incremental improvements in blood sugar starting within the first year.

Their hemoglobin A1c started out at over 8—a sign of uncontrolled diabetes—and ended at about 6.3 on average, with every year resulting in incremental improvement. To put this in perspective, hemoglobin A1c below 5.7 is considered normal, 5.7–6.4 is a sign of prediabetes, and type 2 diabetes is when the A1c is more than 6.5. These men got below 6.5, and they did so by about the third or fourth year. Clearly, testosterone replacement therapy reversed their diabetes, helped them lose weight, reduced their waist circumference, and improved their quality of life.

Another study published in *Obesity Research and Clinical Practice* in 2014 evaluated 255 men ages 33–69 (with the average age being 60) who had gone to a clinic to be evaluated for low testosterone. Some of these men had type 2 diabetes. The diabetic men were seeking urology consultation for erectile dysfunction and decreased libido and were concerned about their testosterone status. They were not offered testosterone as part of the treatment for diabetes from their primary care doctors although they had a testosterone level rate of about 300–360 ng/dL. While this is very low testosterone, most laboratories in the United States would report that level to your doctor as "normal." It really isn't—it's not even close to normal—but that is how we gauge testosterone in the United States. In this study, these men were given long-acting testosterone about every three months.

The researchers in this study were a little bit more enlightened and treated men with low testosterone even though US labs had identified the men as "normal." And the researchers didn't just take the men from a low level to less of a low level—they took them from the 300 ng/dL range up to about 500 ng/dL. Again, this is not quite ideal, but it's getting very close to ideal for most men. The men lost several inches off their waist: a little over an inch in the first year, two inches in the second year, almost three inches in the third year, and over three inches in four years. In the fifth year, they lost three and half inches on average.

They also lost weight every year: starting at one month, the weight started coming off, and they ultimately lost an average of 46 pounds. This is over five years. No men in this group gained weight; all of them lost weight, no exceptions. This is not something we see in the conventional treatment of diabetes, nor do we see that happen with weight-loss drugs. Fasting blood sugar over the five years decreased to a normal range of 97. The hemoglobin A1c went from an average of 7.4 (diabetic) down to an average of 6 (not diabetic).

On average, not only did their blood sugar get better but also their serum cholesterol got better, and their HDL—which is the good cholesterol—rose even as their triglycerides lowered. Their blood pressure improved, and their C-reactive protein (a marker of inflammation) went from an elevated level of 4 down to 0.7, which within is a normal range. In short, testosterone replacement in this group took their A1c from a diabetic range to normal and also reduced their risk of heart disease as a result of their C-reactive protein going from elevated down to normal.

There were no major adverse events related to testosterone replacement. Unlike current conventional treatments for diabetes, the men in this study who were also on true testosterone replacement ("true" as in restoring them to healthy levels) saw their diabetes reversed. No diabetic medication can accomplish that.

## In summary

Diabetes is a global health problem. In men, low testosterone is very common, almost universal in men with type 2 diabetes. TRT has been shown to improve and potentially reverse type 2 diabetes, lead to weight loss, and decrease belly fat. More importantly, TRT reduces mortality in men with type 2 diabetes.

# 6

# Testosterone and Cardiovascular Disease

Als discussed in the introduction, the FDA made an enormous blunder when it required manufacturers to warn physicians that the use of testosterone products could be associated with an increased risk of cardiovascular disease such as heart attacks and strokes. The FDA also specifically wanted manufacturers to alert doctors that testosterone replacement should be used to treat only certain specific conditions that affect the brain or the testicles and to treat men born with an extra female chromosome. The FDA acknowledged that doctors were prescribing testosterone replacement therapy (TRT) for men who had low testosterone levels, but even though those men were seeing overall health benefits, the FDA recommended that testosterone not be used to treat low testosterone due to "normal aging" or when low testosterone occurs in conjunction with other conditions such as diabetes and obesity.

Even though hundreds of scientific papers have shown a clear link between low testosterone and increased cardiovascular disease—and extensive research has shown that testosterone replacement conveys cardiovascular protection—the FDA chose to base its recommendations and rulings on a very small number of studies. One of the main sources of information the FDA used as a basis for its incorrect claim is a paper that was published in the *Journal of American Medical Association* on November 6, 2013 (vol. 310, no. 17). This flawed study was called "Association of Testosterone Therapy with Mortality, Myocardial Infarction, and Stroke in Men with Low Testosterone Levels." The data, among other errors, suggest that men with preexisting heart disease had increased risk of adverse outcomes (stroke or nonfatal heart attacks) if they were put on testosterone replacement. It was later discovered that the methods used were invalid, and the data were corrected to show TRT to be cardioprotective, but it was too late: the incorrect information had already led to a fundamental shift in the FDA labeling rules for testosterone products.

A study published in the January/February 2014 issue of *Canadian Urology Journal* by the Canadian Urological Association looked at the collective knowledge of general practitioners and cardiologists regarding testosterone replacement. Their findings revealed that most physicians—even though they treat cardiac patients—were not aware that testosterone deficiency syndrome has deleterious cardiovascular effects. These were physicians who specialized in cardiac disease as well as physicians who specialized in health and prevention, yet they generally did not know that low testosterone is associated with higher cardiovascular events along with higher rates of diseases that lead to cardiovascular events (i.e., obesity, diabetes, and hypertension; abnormal cardiac lipids are also factors). Other than smoking, testosterone deficiency syndrome is likely *the* most common predictor of subsequent cardiovascular disease, but the doctors who had participated in the study were generally not aware of that. They also weren't aware of the fact that testosterone replacement mitigates much of that risk because it substantially reduces premature death from those conditions.

Back in 2009—well before the FDA rulings in 2014 and 2015—the *European Journal of Endocrinology* published a study that evaluated men with known cardiac disease as evidenced by abnormal stress tests and diagnostic imaging (with the assistance of the latter, they looked at the thickness of the men's arterial blockages). When these men were put on long-term testosterone replacement, they had less myocardial ischemia, less angina, and more exercise capacity. They also had some degree of reversal of their atherosclerosis as evidence of measurement of plaques in their arteries.

## International Expert Consensus Resolutions on the FDA's ruling

We talked about the International Expert Consensus Resolutions in the introduction, but I still want to go over one of the resolutions here. It's from the July 2016 resolution "Fundamental Concepts Regarding Testosterone Deficiency and Treatment: International Expert Consensus Resolutions." These resolutions are based on the joint conclusions of a large group of expert physicians who treat men's health; these experts came from of a broad range of medical specialties across 11 countries and four continents. All of the resolutions the experts issued had been unanimously approved by the members. The FDA was invited to the meeting but did not attend, which is not surprising.

*Consensus Resolution 7: "The evidence does not support increased risks of cardiovascular events with testosterone therapy."*
After extensively studying the medical literature, the experts had five specific comments regarding this resolution, and they also had a comprehensive explanation. Let's review their five comments.

1. "Two observational studies received intense media attention after reporting increased cardiovascular risks. Both had major

flaws/limitations. One misreported results; the other had no control group.

2.  "Low serum testosterone is associated with increased atherosclerosis, coronary artery disease, obesity, diabetes, and mortality.

3.  "Several randomized controlled trials in men with known heart disease (angina, heart failure) show greater benefits with testosterone versus placebo (greater time to ischemia, greater exercise capacity).

4.  "The largest meta-analysis showed no increased risk with testosterone therapy: *reduced* risk was noted in men with metabolic conditions.

5.  "No increased risk of venothrombotic events with testosterone therapy."

(Note: Venothrombotic events refer to things such as blood clots in the leg or lung.)

## Let's talk about the science

In a previous chapter, we reviewed the study of more than 80,000 men in their 60s with low testosterone followed for about six years. The men who were treated with TRT were half as likely to have a heart attack or stroke or to die. Period. Other studies reaffirm the same findings.

A study in *The Aging Male* that was published in 2015 from the United Kingdom began in 1989 and went on for 25 years. The goal was to determine whether the long-term benefits of testosterone replacement continued over time. In terms of studies, 25 years is an amazingly long period of time to study. It's worth nothing that at the beginning of the study, the physicians generally had not checked the men's testosterone levels because the results often aren't very accurate and physicians know that such results do not always correlate with symptoms of low testosterone. The physicians based the decision of whether to treat with testosterone on symptoms of testosterone deficiency, a generally accepted practice that has been going on for at least 70 years.

Another study published in the *Lancet* (one of the most respected medical journals in history) was conducted by a group of health systems in Toronto, Canada. The study included 10,311 men who were treated with testosterone replacement therapy and compared them to untreated matched controls numbering 28,029. The study was conducted from January 1, 2007, to June 30, 2012.

This study tracked men for an average of 5.3 years. What the researchers found was quite astonishing: the men with the lowest amount of testosterone exposure had the highest risk of mortality in cardiovascular events compared to the controls. In other words, the men with low testosterone who took testosterone for a very short time or who never took testosterone at all had the highest rate of death and heart attacks. That means that *not* taking testosterone replacement therapy when you have low testosterone doubles your risk of dying from anything, including heart attacks. If you *do* take testosterone therapy, the longer you take it, the more cardiovascular protection you have. And not only did the men in this study who used TRT for the longest amount of time have the lowest risk of dying and heart attacks but also they had the lowest rates of prostate cancer.

For years, there's been compelling evidence that low testosterone predicts increased cardiovascular events. A study done in *Hypertension Medical Journal* from the Johns Hopkins University School of Medicine analyzed carotid artery health and compared it to various hormones, not just testosterone. The reason for assessing the carotid artery in particular is because it's right in the neck, very close to the skin. (This is where you put your fingers on your neck to feel your pulse.) If the lining is thin, the cardiovascular system is hearty and healthy.

The study in *Hypertension Medical Journal* looked at specific hormones, such as estradiol, testosterone, and other sex hormones; results were adjusted for the presence of conditions such as diabetes, smoking, blood pressure medications, and lipid levels. Still, even when all of those adjustments were made, the researchers found that men with higher testosterone levels had healthier carotid arteries and men with lower testosterone levels had more unhealthy carotid arteries. This is physical evidence of what we see with observational studies. (In this case, an observational study is where we notice that men with low testosterone have more cardiovascular disease and men with normal testosterone have the least cardiovascular disease.) This study explained why a low-testosterone environment leads to more atherosclerosis . . . which was predictable because we already know that low testosterone leads to obesity and diabetes and hypertension, all of which are linked to higher incidents of atherosclerosis.

Another study sought to measure a different kind of vascular risk using a different method. Again, an observational study allows us to follow men for a long time and see whether they have heart attacks and then compare that factor to some other factor. We can also look at specific physical findings associated with cardiovascular disease to determine what conditions increase the risk of cardiovascular disease.

In 2015, the Hypertension in Africa Research Team located in South Africa published a study that measured the retina, which is an inner coating of the eyeball laced with tiny blood vessels. These tiny blood vessels often become diseased when people have untreated hypertension (i.e., where cardiovascular disease is present). The researchers compared the retinal microvascular level of health to testosterone levels. As you could expect, men with a normal testosterone level had healthy retinal microvasculature, whereas men with low testosterone had unhealthy retinal microvasculature.

## Evaluating cardiovascular risks and optimal treatment

It is pretty well established that low testosterone leads to more cardiovascular disease and that testosterone replacement substantially reduces the risk of cardiovascular events and reduces the risk of premature death by about 50% in men in their 60s. It's also pretty well established that this protection continues as men reach their 70s and 80s. Since cardiovascular events are rarer in men under 50, we might want to look at other features of cardiovascular risk to determine the relative protection testosterone offers. We know that testosterone replacement is cardioprotective—that is, there is evidence that it reduces deaths, cardiovascular events, and physical evidence of cardiovascular disease—so shouldn't scientists start evaluating what aspects of cardiovascular risk that testosterone replacement would have the most impact on?

A study published in 2014 by the International Society for Sexual Medicine in its journal, *Journal of Sexual Medicine*, evaluated the long-term treatment of men with low testosterone who received testosterone replacement therapy. This five-year-long study looked at fasting blood sugar, weight, and diabetes, all of which are independent risk factors for heart disease. (I talk about these factors in the diabetes and weight chapter.) This study also looked at total cholesterol, LDL cholesterol, HDL cholesterol, triglycerides, and blood pressure. Remember that LDL cholesterol is the bad cholesterol, and HDL cholesterol is the good cholesterol.

What the researchers found in this five-year study was that cholesterol drops from high to high normal (about a 15% drop in the first year) and then continues to drop every year after that even though these men were *not* put on cholesterol-lowering medications. The LDL dropped precipitously during the first year and continued to drop—albeit not as sharply—for the next five years. The HDL went from low normal to high normal. This is good, because HDL is the good cholesterol—there is no upper limit. Although the total cholesterol level dropped only about 15% at first, that value measures both good and bad cholesterol, so it's not a completely accurate picture of the overall cholesterol situation. If both the HDL and LDL factors were fully taken into account, the numbers would show about 25% improvement. Triglycerides are also a concern when assessing serum lipids. These are free fats that flow through our bodies and stick to our arteries, and we don't want much of them. Again, the triglycerides dropped precipitously in the first year and continued to slowly drop every year thereafter.

Another study done by the same group (but published separately) looked at HDL and LDL cholesterol in men who took testosterone replacement therapy for about five years, stopped taking it for years for various reasons, and then resumed taking TRT. The researchers compared the changes in their LDL and HDL cholesterol to changes in their testosterone levels during that same time period. The purpose of a study like this was to determine whether the testosterone replacement cured a problem and then was suddenly no longer needed or whether TRT should be taken for the remainder of a man's life.

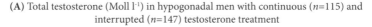

(A) Total testosterone (Moll l⁻¹) in hypogonadal men with continuous (*n*=115) and interrupted (*n*=147) testosterone treatment

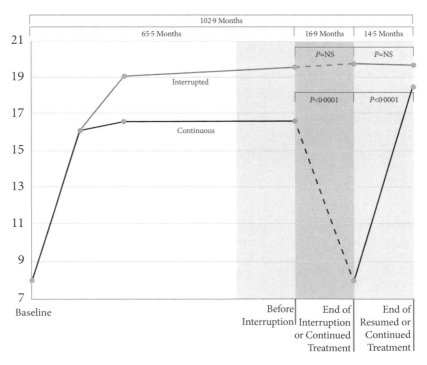

This graph shows that the men represented by the top line saw improvement of their testosterone levels over the five years of the study. When they stopped taking TRT, their level went back to their initial baseline. When they resumed testosterone, they went back up to replacement levels. The bottom line represents men who were compliant and who took testosterone for about 8.5 years. It's interesting to note that the men who were corrected to a level of about 16 nmol/L (that's 450 ng/dL; this was a European study, so researchers used nmol/L rather than ng/dL) were more likely to be noncompliant than the men who were corrected to a level of 19 nmol/L (or 550 ng/dL). As we've previously talked about, more than 50% of men experience symptoms of low testosterone when they have a blood level rate of around 450 ng/dL or less. Very few

(C) LDL cholesterol (mg dl⁻¹) in hypogonadal men with continuous (*n*=115) and interrupted (*n*=147) testosterone treatment

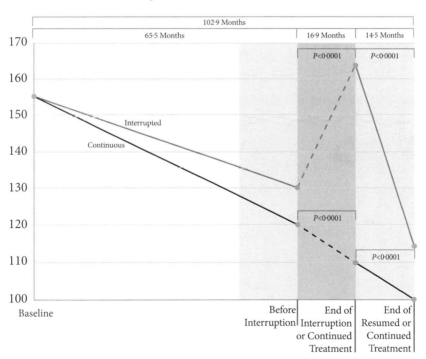

men with a level of 550 ng/dL have symptoms of low testosterone, so it's understandable why the partially treated men might have been noncompliant—they received some benefit from TRT but not as substantial as the benefits the men who were treated more appropriately experienced.

Now, let's look at their lipids.

The men's LDL at the onset of the study was on the high side. As you can see, the men represented by the top line saw more improvements in their LDL levels because they were treated more appropriately with testosterone levels (i.e., they wound up in a more ideal range). The men represented by the bottom line still benefitted, but not as much. When they discontinued testosterone, as expected, their LDL went back up; when they resumed testosterone, it went back down. These results show a direct cause and effect.

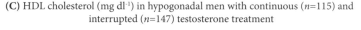
**(C)** HDL cholesterol (mg dl⁻¹) in hypogonadal men with continuous ($n$=115) and interrupted ($n$=147) testosterone treatment

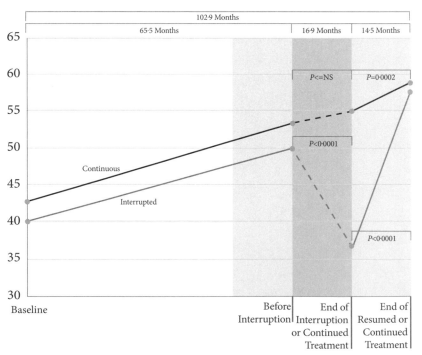

Their HDL cholesterol, or the good cholesterol, shows a similar pattern: when the men took testosterone replacement, it went up; when they stopped testosterone, their HDL went down; when testosterone was resumed, HDL went back up.

Another marker of future cardiovascular disease is called CRP, or C-reactive protein. If your CRP is high, your risk of cardiovascular events is high; if your CRP is low, your risk of cardiovascular events is low. This is about 40% more accurate in predicting cardiovascular disease than LDL cholesterol is, which is why CRP levels are commonly checked as well. So what happens to CRP when testosterone is initiated, stopped, and then resumed? Let's take a look.

Their CRP follows the same pattern that their LDL cholesterol took: when they were treated with testosterone, their CRP went down, and

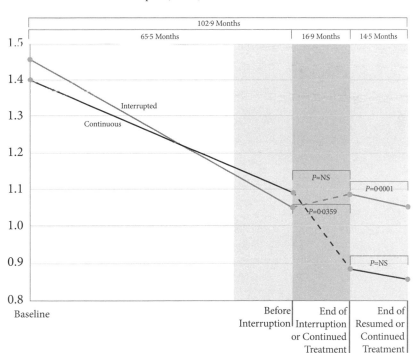

(C) CRP (mg/dl) in hypogonadal men with continuous (n=115) and interrupted (n=147) testosterone treatment

then when they discontinued testosterone, their CRP went back up; when they resumed testosterone, it went back down.

And what about blood pressure? Hypertension is clearly a risk factor for cardiovascular disease, so what happens when men with low testosterone who also have blood pressure take TRT? Again, measurements were taken on men who took testosterone replacement for several years, stopped taking it for years, and then resumed it.

Blood pressure is made up of two different numbers: systolic blood pressure (the top number) measures the force that your heart exerts on your blood system when it is contracting, and diastolic blood pressure (the bottom number) measures the force exerted between blood vessels when the heart is at rest and refilling.

**(A)** Systolic blood pressure (mmHg) in hypogonadal men with continuous (*n*=115) and interrupted (*n*=147) testosterone treatment

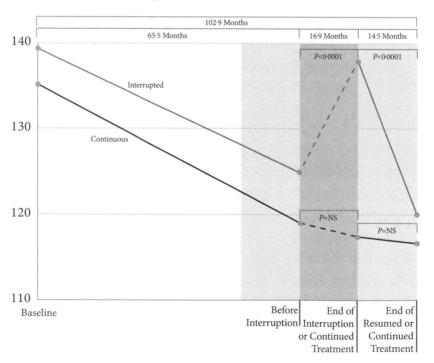

If the systolic number is less than 120 mmHg, that indicates normal blood pressure. Prehypertension is defined as being between 120 and 139, and hypertension or high blood pressure is over 140. Diastolic blood pressure is normal when it's less than 80, whereas a range of 80–89 indicates prehypertension. A number over 90 means hypertension.

In these graphs, you can see that the men went from prehypertensive levels to normal levels with testosterone replacement. When they stopped taking testosterone, their blood pressure crept back up; when they resumed testosterone, their blood pressure improved again. It's interesting to note that the men saw an improvement every year in spite of the fact that men's blood pressure normally gets worse with age. These men, however, showed an improvement in blood pressure as they got older.

**(C)** HDL cholesterol (mg dl⁻¹) in hypogonadal men with continuous (*n*=115) and interrupted (*n*=147) testosterone treatment

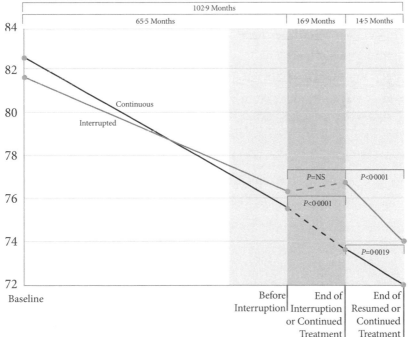

Again and again, the same pattern occurs: testosterone replacement in men with low testosterone is like a switch that you turn on to get healthier. When you turn it off, you become unhealthier, generally the way you were before you took testosterone. Then, if you resume taking it, you start getting healthy again. We can also see from this study that the relative level of testosterone replacement determines the relative outcomes of beneficial cardiovascular markers.

A study published in *Diabetes* in May of 2013 sought to remove other factors that may contribute to the beneficial effects of testosterone and cardiovascular protection. For this study, researchers recruited healthy young men with normal testosterone. They then gave the men medications to completely block testosterone (which is a means of medical castration). In other words, the men were given medication to make their testicles stop working. Who would volunteer for this kind of study? I've no idea, but the researchers did find several men willing to participate.

In summary, what the researchers found was that by chemically castrating the men, their testosterone levels dropped precipitously. This led to oxidation of their cardiac lipids, which is to say that the process of acquiring low testosterone made the men's cardiac lipids very inflammatory, harmful, and dangerous. There were no other variables in this study—these were healthy men with normal testosterone levels who were made to suddenly have low testosterone. As soon as their testosterone dropped, their cardiac lipids became their enemy. When their testosterone levels resumed and went back up to normal, the men essentially replaced all of their cardiac lipids with noninflammatory ones.

An article was published in the *Heart Journal* in 2003 that evaluated the response of men with myocardial ischemia—also known as angina—to TRT. (Their angina was proven by a treadmill stress test.) These men presented with very low testosterone levels and were given testosterone replacement. The men responded with improved stress test results. Several studies have looked at what happens when a man with low testosterone—in this case, extremely low—takes TRT and raises his level to a not-so-extremely-low level. Even when their levels are raised a little bit, these men still see improved stress test results.

## What's happening on the cellular level?

We've talked about the fact that when men are put on testosterone replacement therapy, their risk factors for cardiovascular disease improve: for example, there's less obesity, less hypertension, better cardiac lipids, and improvements in diabetic markers.

But what's happening at the cellular level? A study published in the *Translational Andrology and Urology* in 2016 evaluated middle-aged men with low testosterone. This study measured endothelial function. That refers to the relative health of the arteries, which are the soul, so to speak, of the cardiovascular system. This function is measured and watched over time to see whether people get worse or better.

The researchers measured baseline endothelial function in the men who were on testosterone replacement therapy and then measured them again at six months. In that short period of time, the men's endothelial function improved by 25%. There really are no drugs available that give us this kind of improvement.

Let's revisit the study with the 83,000 men who were followed by the VA for several years. That study showed that the men with low testosterone who were given testosterone replacement were half as likely to die and half as likely to have a heart attack or stroke compared to the men who did not receive testosterone replacement.

What would happen if we followed men on testosterone replacement even longer? Would studies done over even longer periods of time reveal better protection? Since men are more likely to develop vascular disease as they age, does this protection stay in place as men age but continue to be on testosterone replacement versus those who choose not to replace testosterone?

A study was published in the *Journal of Cardiovascular Pharmacology and Therapeutics* in February of 2017 regarding men with low testosterone.

When the study began, the men had an average age of 61. The researchers compared the cardiovascular outcomes of the men who took testosterone replacement to those who did not. When they evaluated the reasons why some of the men with low testosterone did *not* seek treatment, the men said their most common reason was fear about health concerns, particularly about the cardiovascular risk that the FDA requires drug manufacturers to list on the labels of testosterone replacement products.

This study followed the men for about eight years, which is a long time for a clinical study. Because the results of this study were so profound and were the antithesis of what the FDA has ruled, I wrote about this study in a blog post. Here's what happened:

> *There were 656 men. About half received testosterone replacement while the other half chose not to take TRT for fear of health concerns. At the end of the eight years, there were 75 cardiovascular events, which included heart attacks and stroke.* **All 75 cardiovascular events occurred in the group of men who choose not to take testosterone. The men who choose to take testosterone experienced 0 cardiovascular events.**

Of the men who chose not to take testosterone, there were 21 deaths, 19 of which were related to cardiovascular events such as heart attack and stroke.

In the group of men who took testosterone, there were two deaths unrelated to cardiovascular disease at the end of eight years, both from natural causes unrelated to cardiovascular disease. In the group of men who chose not to take testosterone because of misguided fears of health concerns, 21 men died, 30 had nonfatal strokes, and 26 had nonfatal heart attacks. But that's not all.

## Blood sugar and diabetes

The men with testosterone replacement saw favorable improvements in their blood sugar, while the men in the control group—that is, men who did not take testosterone—saw no improvement in their blood sugar. More profound yet were the changes in hemoglobin A1c. (Remember that hemoglobin A1c is a marker of diabetes.) The men who took testosterone replacement had a substantial improvement of hemoglobin A1c, meaning they were much less likely to develop diabetes. However, hemoglobin A1c worsened over the next several years in the men who chose not to be treated with testosterone replacement.

## Blood pressure

Both systolic and diastolic blood pressure improved over the observed period. In the men who took testosterone replacement, their blood pressure went from elevated to normal, while the men who opted out of testosterone replacement therapy saw their blood pressure worsen over the same period.

## Cholesterol

The study also measured cardiac lipids. The men who took testosterone replacement had significant improvements in their cardiac lipids, while the men who did not take testosterone replacement saw a typical worsening of their cardiac lipids over time.

## Liver function

The men who took testosterone replacement therapy also saw improvements in the health of their liver, which is the body's toxin-removing organ. The men who did not take testosterone replacement saw worsening liver function over time.

### Weight and obesity

The men who took testosterone replacement had a significant and sustained weight loss over the observational period: every single year during the eight-year follow-up period, the men's weight improved when they took testosterone replacement. Their average total weight loss was about 40 pounds. In contrast, the men who did not take testosterone replacement steadily got heavier over the following years.

### Belly fat

The men with testosterone replacement saw a progressive improvement of their waist circumference, while (of course) the men who did not take testosterone steadily got thicker around the waist.

### Prostate cancer

In the follow-up group, men with testosterone replacement had 50% fewer cases of prostate cancer compared to the men who did not take testosterone replacement. This finding aligns with numerous studies that have shown that testosterone replacement does not increase the risk of prostate cancer—in fact, it may be protective.

### Heart attack and stroke

In the men who were treated with testosterone replacement, there were no cardiovascular events such as heart attack or stroke. Let's repeat that: there were *no* deaths from cardiovascular disease. During the eight years of the study, there were two deaths in the group of older men, but those deaths were unrelated to cardiovascular disease.

In the men who did not take testosterone replacement, there were 21 deaths, with 19 of them related to cardiovascular events such as heart attack and stroke. There were an additional 30 nonfatal strokes and 26 nonfatal heart attacks. Again, this compares to *zero* in the group of men who took testosterone replacement.

It's true that even though none of these men in this study who took testosterone had a heart attack or stroke (versus the high occurrence of those incidents among men who did not take testosterone replacement), testosterone replacement does not *guarantee* that a heart attack or stroke won't happen, nor is it a guarantee against prostate cancer. However, we can see that men are *less likely* to get prostate cancer, have a heart attack or stroke, or die prematurely when testosterone is taken.

As we noticed earlier in this study, there was a 50% reduction in cardiovascular events and premature death. This study shows substantially more divergent outcomes between the treated and untreated groups because the study lasted for a longer period of time—the longer you don't take testosterone when you have low testosterone, the more your odds of having adverse cardiovascular events and premature death go up. Over the long run, that equation creates a bigger gap in health outcomes between the men who seek treatment and those who avoid it.

## Connections to blood clots and congestive heart failure

To wrap up this chapter on cardiovascular disease and testosterone, I want to talk about two separate subjects: the connections between TRT and blood clots and the connections between TRT and congestive heart failure.

## Blood clots

Venous thromboembolism (VTE) is a health condition where blood abnormally clots inside veins. When this occurs in the legs, it's called DVT, or deep venous thrombosis. In the lungs, it's called a pulmonary embolism, or PE.

These are both very potentially serious conditions. A blood clot in the legs can lead to swelling and long-term pain, but, more importantly, it could break loose and travel to the lungs, causing a pulmonary embolism. This can be fatal.

Most blood clots are caused by immobility and injury with or without genetic predisposition—for example, clots often happen in patients who are hospitalized for various conditions. Because of the frequency of clots forming in hospitalized patients or in postsurgical patients who are immobile, anticoagulants are often used in those settings.

The FDA has required drug manufacturers to label their products with a warning that all testosterone products can increase the risk of blood clots. The agency says this warning is justified because some men have had blood clots while taking testosterone. There is no scientific basis for this rule by the FDA, but it still stands.

An article published in *Chest* (a major medical publication that focuses on blood clot issues) sought to evaluate this matter definitively. In their study, researchers reviewed the records of 71,407 men and compared the risk of blood clots in men who received testosterone to those who did not. The researchers also further broke down the results to evaluate relative testosterone blood levels, separating the men who achieved satisfactory testosterone levels from those who did not.

In the study of more than 70,000 men, about 38,000 had low testosterone, were given testosterone replacement, and subsequently had blood tests showing that the testosterone was adequately replaced. The second group was made up of about 22,000 men with low testosterone who were given testosterone replacement. Follow-up blood tests showed they were not taking an adequate amount of testosterone to give them a normal blood level. The third group was made up of men with low testosterone who were never treated with testosterone replacement; these men continued to have low testosterone.

In this study, researchers found no relationship between the rate of blood clot development in treated and untreated men with respect to testosterone levels no matter whether they were treated with testosterone replacement. There was simply no link.

Let's go over a study that addresses the more intricate aspects of blot clots. A study published in *Urology* in August of 2015 evaluated men with low testosterone, evaluated whether they were put on testosterone replacement therapy, and looked at the rates of myocardial infarction caused by a blood clot, strokes caused by a blood clot, blood clots in the legs (DVT), and blood clots in a lung (PE). What the researchers found is that testosterone replacement therapy did *not* cause myocardial infarction, strokes, PE, or DVT; rather, the testosterone replacement therapy was protective. The researchers also found that men with low testosterone who were *not* treated with replacement therapy were more likely to die of any cause, particularly of blood clot causes. The assumption that testosterone increases the risk of blood clot is wrong—it is, in fact, the opposite.

## Congestive heart failure

Now let's talk about the final cardiovascular topic: congestive heart failure. Congestive heart failure is a condition where the heart is damaged and the muscles are fatigued and dying. This leads to an inability of the heart to fully circulate blood throughout the body. There are several forms and causes of congestive failure, but let's talk about the main cause, which is cardiovascular disease. (Congestive heart failure is also associated with diabetes, obesity, hypertension, and atherosclerosis, but all of those conditions are frequently present when men have low testosterone and/or cardiovascular disease.) Congestive heart failure is more common in men with low testosterone.

My father died at a young age of congestive heart failure. He was one of seven brothers, and they all had heart disease at a very young age and had heart attacks in their 40s. Like my father, they all died pretty young. They also all had large waistlines and were overweight. I'm not sure whether my father's brothers had diabetes—most of them died before I was born or when I was very young—but my father had diabetes. I'm sure he had low testosterone, too, but I was not practicing in that field of medicine then and didn't know anything about it, and most doctors were generally unaware of testosterone deficiency at that time. My father had bypass heart surgery in his 40s, had a couple strokes, and became a cardiac cripple at a young age. He developed heart failure, had chronic chest pain, was always short of breath, and had very little exercise tolerance. All of this had happened while I was still at a young age. He really couldn't do anything active—I remember that we would always park as close to the door as possible because it was hard for him to walk. He always avoided physical activities. I wish I knew then what I know now.

A study published in the *Journal of Thoracic Disease* in 2016 reviewed two clinically published trials to determine whether testosterone replacement therapy benefits patients with congestive heart failure. What was the result? The researchers found that testosterone replacement improved exercise capacity, muscle strength, and EKG readings. Exercise capacity is measured by simple things such as the ability to walk, and for people suffering with congestive heart failure, improving their ability to walk is a substantial benefit. TRT did not cure heart failure in the men in the trials, but it did improve their quality of life.

Other studies have shown improvements in simple things like handgrip strength, walking distance, and physical functioning capacity. Studies have also shown that testosterone replacement improves insulin sensitivity and decreases the symptoms of heart failure.

## In summary

Testosterone replacement therapy has been shown to improve carotid blood flow (i.e., blood flow to the brain). Not only has it been shown that TRT reduces the progression of atherosclerosis but also it can actually—at least, to some degree—reverse atherosclerosis. Testosterone has been shown to lower blood pressure from the prehypertension range to a normal range. Testosterone replacement has been shown to reverse or improve many of the risk factors for cardiovascular disease, such as obesity, waist circumference, diabetes, and metabolic syndrome. Testosterone is associated with improved cardiac lipids and lowered cardiovascular inflammation.

Testosterone deficiency is clearly linked to increased risk of all-cause mortality—in other words, premature death. Low testosterone is clearly associated with premature cardiovascular disease.

In large-scale published clinical trials, testosterone replacement therapy has been shown to reduce the mortality rate from any cause by over 50%, and regardless of a man's age, TRT has been shown to be beneficial to the cardiovascular system when men have low testosterone. The current bias that testosterone replacement should occur only when a man's testosterone is extremely low is not supported by the literature.

The FDA ruling that testosterone replacement therapy be used only in men who have severely low testosterone caused by organ damage or an extra female chromosome is absurd, but nonetheless, the FDA has reaffirmed its position regarding the labeling of testosterone medication—the agency continues to require manufacturers to warn patients and doctors that testosterone might cause an increased risk of cardiovascular disease and stroke. The FDA has acknowledged that there is no evidence to support its conclusion, but still, the agency has continued to cite the two flawed medical journal articles as a reason to keep this incorrect warning in place.

Health insurance companies—not just in the United States but also elsewhere—often require verification of extremely low testosterone on consecutive days at a time of day when testosterone levels are naturally at their peak. This is obviously required because insurance companies are trying to avoid having to pay for medications that benefit people even though the companies have already accepted people's insurance premium payments. There is no scientific rationale for these requirements—numerous studies have shown that there is very little correlation between absolute blood testosterone levels and symptoms of low testosterone. There is also very little correlation between how well absolute testosterone levels determine a man's potential response to TRT. This has been confirmed in numerous clinical studies (both observational and controlled trials). Also, numerous medical organizations and expert consensus panels have refuted the practice of relying on absolute blood testosterone to determine whether treatment is given.

In brief:

- Testosterone replacement therapy protects against not only cardiovascular mortality but also all-cause mortality.

- Testosterone replacement therapy does not increase your risk of heart attack or stroke; it decreases it.

- Testosterone replacement therapy reduces symptoms associated with congestive heart failure, which is the endpoint of cardiovascular disease.

- Testosterone replacement therapy does not increase your risk of blood clots; it decreases it.

- Testosterone benefits your cardiac lipids, reduces your risk factors for cardiovascular disease, reduces your risk of developing cardiovascular disease, reduces your risk of dying from a heart attack or stroke, and reduces your chance of dying from any cause.

And all of this doesn't even get into why men seek out testosterone replacement therapy in the first place, which is generally because they're concerned about weight gain, sexual dysfunction, and/or declining energy levels. I think this is a big gaping hole in health care today. We have irrefutable evidence that testosterone replacement therapy is protective against obesity and can help men lose weight. Testosterone replacement therapy also has beneficial effects on diabetes—namely, it appears to reduce and reverse diabetes. Likewise, testosterone replacement therapy reverses features of metabolic syndrome: it improves blood pressure and cardiac lipids and is cardioprotective. When taken long-term, even inappropriately low testosterone replacement therapy improves men's survival when they have low testosterone. All of these beneficial effects have been proven and are well beyond the point of controversy.

The debate about TRT is completely fabricated. The FDA felt pressured to reassess its requirements for the labeling of testosterone replacement therapy; physicians and healthcare professionals are biased against testosterone replacement; insurance companies do not want to pay for it . . . and so a controversy was created. But physicians who study medical literature and practice evidence-based medicine—and who lack personal bias—are well aware of the actual, proven assessments of the effectiveness of TRT.

Most men seeking testosterone replacement therapy aren't doing so for the purpose of cardiac protection, nor are they thinking about TRT to manage their diabetes, improve their metabolic syndrome status, or lose weight (although TRT does have all of those benefits). Most men seeking out testosterone replacement therapy are doing so because they want to feel better and restore lost their vigor, libido, and energy. The "Low T" campaign carried out by drugmakers was a result of the companies' wanting to make consumers aware of their options so they could talk about them with their doctors. As we've previously discussed, these campaigns were (of course) shut down by the FDA, so we won't be hearing any of those campaigns anymore.

Still, as a society we adapt to what studies and evidence show. I wrote this book. I do speaking engagements with other physicians. I write blogs. Other doctors produce medical research, band together with other medical professionals, or create expert consensus opinions or medical society position papers. The FDA ruling will not shut us down. The FDA will not prevent you from maximizing your health.

# 7

# Testosterone and Urinary Tract Health

In this chapter, we'll cover the myths and the realities surrounding low testosterone and urinary tract health as well as the myths and realities surrounding testosterone replacement and urinary tract health.

It is a common misconception that testosterone replacement may increase the risk of prostate cancer. There is no evidence for this whatsoever. Extensive research has been done and totally disproves this. Some also believe that testosterone replacement can make men develop urinary symptoms. This is also incorrect. In general, urinary health improves with normalization of testosterone levels by testosterone replacement therapy.

One of the things we're going to talk about is the PSA (prostate-specific antigen) blood test, which is a test used as an aid for detecting prostate cancer. PSA results can potentially be affected by initiation of testosterone replacement therapy. There are many misconceptions regarding the PSA blood test, so I'm going to clarify those for you. First, we'll review symptoms that are commonly associated with prostate hypertrophy; we'll call these urinary tract symptoms.

There remains some misconception in health care that perhaps testosterone replacement leads to more prostate cancers. This is totally false and has been proven to be false time and time again. One of the reasons for this misconception is that when men do have prostate cancer, certain types of it will respond to drugs that lower the testosterone level to about zero. There are also studies that have looked at treating prostate cancer with supraphysiological doses of testosterone, which means testosterone levels that exceed normal. Note that testosterone does not cause prostate cancer—prostate cancer occurs in the presence of normal or low testosterone. There is some evidence that men with low testosterone have slightly more prostate cancer, but this is not definitive. And testosterone replacement therapy is generally not associated with reduced rates of prostate cancer. But in answer to the question of whether TRT causes prostate cancer, the response is: No! Testosterone replacement therapy does not increase the risk of developing prostate cancer.

## The PSA blood test

Since the PSA blood test has been used for years as a tool to help detect prostate cancer, I'm sure you've heard of this test. (In fact, if you're over 50, your doctor has probably ordered it for you.) Let's look at the current guidelines and concepts behind the test.

A normal result is less than 4.0 ng/mL. There's also a test value called the "velocity of change," which means that there is substantial change in the test results over a short period of time. The velocity of change does not appear to be clinically meaningful, so we mostly look at the absolute PSA level.

PSA testing is generally initiated when men turn 50 (maybe a little bit younger for men who are at higher risk) or perhaps when men turn 45 (such as in the case of African American men, who typically have a higher risk of prostate cancer). Currently, the recommendations from urological societies are that before you get a PSA blood test, you give your **informed consent about the risks of the blood test**. This might seem unusual—why would they recommend that a patient be informed about the risk of doing a blood test?

Because of something we call "overdiagnosis." Prostate cancer is relatively common—if men live long enough, they will eventually get prostate cancer (which can be determined by a biopsy). It is extremely unlikely that prostate cancer will kill them, because at that point, they are very old and will probably die from something else. Prostate cancer grows very, very slowly. It's true that a young man with prostate cancer has a reasonable chance of the disease progressing to the point that he could potentially die from it. But when prostate cancer is detected when a man is over 70 years old and it takes decades to spread, it is unlikely the man will be alive to die from prostate cancer—he'll die from something else before then.

A study called the Prostate, Lung, Colorectal, and Ovarian Cancer (PLCO) Screening Trial was carried out to determine whether screening for cancers saves lives. In the PLCO Trial, the prostate detection was done by a PSA blood test and a digital rectal exam.

What the researchers found is interesting. The men who had the annual prostate cancer screening had a higher incidence of prostate cancer than the men who were not screened. This is probably because although the men who were not screened potentially had prostate cancer, they didn't know they had it because it didn't make them sick. What's more, in the PLCO Trial, both the men who had regular screening and the men who had no screening had the same rates of deaths from prostate cancer. The digital rectal exam and the PSA did not alter the likelihood of death from prostate cancer. Testing did, however, increase the detection of prostate cancer; it probably also created a lot of fear and concern even though it didn't save lives. Testing certainly led to a bunch of additional surgeries to treat the prostate cancer that was detected, but, again, no lives were saved. To summarize this study, while prostate detection helped detect prostate cancer, it didn't positively alter outcomes—unnecessary testing just led to more (ineffective) medical intervention. This trial went on for 13 years, so the results are meaningful.

Another study called the European Randomized Study of Screening for Prostate Cancer (ERSPC) evaluated cancer deaths among men who were screened with PSA versus men who were not screened with PSA. In this study, the men who were screened had more cancers detected, and they also had a slightly lower death rate from prostate cancer. However, both the men who were screened and the men who were not screened had the exact same mortality rates. Why? As a result of doing unnecessary surgeries on the prostate for a cancer that probably wouldn't kill them, there were more complications in the men who were screened. The two groups wound up having the same mortality rates.

On July 17, 2012, the US Preventative Services Task Force Recommendation Statement evaluated the benefit-versus-harm evidence of PSA testing for prostate cancer detection. The task force determined the following:

> *The USPSTF recommends against PSA-based screening for prostate cancer (grade D recommendation). This recommendation applies to men in the general US population, regardless of age. This recommendation does not include the use of the PSA test for surveillance after diagnosis or treatment of prostate cancer; the use of the PSA test for this indication is outside the scope of the USPSTF.*

This is what the task force found after reviewing evidence based on hundreds of thousands of men who either had regular prostate cancer screening or no prostate cancer screening.

For every 1,000 men between the ages of 55 and 69 who are screened every one to four years over the course of a decade, the task force found there would be:

- 1,000 men screened for PSA for 10 years

- Five men will die of prostate cancer **with no** screening

- Four or five men will die of prostate cancer **with** screening

- **Fewer than one man** will benefit (by not dying) from screening 1,000 men for 10 years

- 100–120 men will suffer moderate to major side effects from unnecessary biopsies

- 110 men will undergo treatment for prostate cancer even though if it was missed, they would have been asymptomatic for life

- Two will have serious cardiovascular events from treatment

- One will develop a blood clot from treatment

- 29 will develop erectile dysfunction from treatment

- 18 will develop urinary incontinence from treatment

- About one will die from treatment, about the same as would survive because of screening

The full details are listed at www.cancer.gov/types/prostate/psa-fact-sheet#r5.

The PSA is controversial in the sense that most doctors are still doing it (and most patients expect it) even though the science does not support it. Case in point: Medicare, a government agency that provides health insurance, pays for an annual PSA test for patients who are over 50 years old. But the US Preventative Service Task Force has reviewed the literature from two large-scale studies (more than 250,000 men) and has determined that not only does the PSA test not do anything to save lives but also it potentially has a negative impact on men's health.

The urologic community and the primary care community are currently trying to figure out what to do with this information. It's pretty new. Most likely, unless your doctor is specifically interested in the subject, he or she may know nothing more about this test than "it's not so reliable." But reality is that the PSA test is probably useless.

Most guidelines for testosterone replacement recommend that we check the PSA annually. I'm pretty torn on this, because obviously I am opposed to doing anything that causes potential harm. In the studies done on whether PSA testing is validated, a majority of the men studied were not on testosterone replacement, and there is no good, solid evidence that checking the PSA when testosterone is replaced would be beneficial. From the standpoint of evidence-based medicine, at this time, it does not make sense to require men taking TRT to do annual PSA testing. Because of this, we do not require it in our practice, but we do offer PSA testing to men as an option and advise them that doing the test is ultimately their choice. We also advise patients that the test appears to cause more harm than good, although we are not absolutely certain of that.

Let's talk about some studies that look at PSA levels in relationship to testosterone replacement.

A study done in the *British Journal of Urology International* in 2013 evaluated long-term use of testosterone replacement over six years and also evaluated follow-up PSA levels. The study was carried out between January 2002 and December 2008. The men in the study had prostate cancer screening, including a digital rectal exam and the PSA blood test. The researchers repeated the blood test every three months during the first year and then every six months thereafter. That's pretty aggressive screening and not recommended for routine detection of prostate cancer.

In this study, the researchers found that there was no relevant change in the PSA level, velocity of change, or significant prostate risk from six-year-long use of testosterone replacement. The intent of this study was put to rest the idea that testosterone replacement causes prostate cancer or even causes any alterations in the screening for prostate cancer.

A study published in December 2015 in the same journal evaluated men who were on testosterone replacement therapy for low testosterone and who had prostate biopsies. What the researchers found was that while men with low testosterone who are treated with testosterone replacement therapy can certainly develop prostate cancer, the researchers did not find any correlation between an initial rise of the PSA upon initiation of testosterone replacement and an abnormal biopsy. I'd like to summarize this study as well as the prior one by saying that, occasionally, men will see a rise in their PSA (although not very much) once they've initiated testosterone replacement. The long-term study showed that over time there are no significant changes but that occasionally a man will have a slight increase in his PSA after initiating testosterone therapy.

Some physicians recommend measuring the PSA when a man starts testosterone. This is your choice. As we discussed earlier, the PSA screening does not save lives—in fact, it leads to more harm than good. That's because the PSA screening leads to more biopsies, which have unnecessary side effects and risks. As we will discuss this in this chapter, the evidence shows us that testosterone replacement does *not* increase the incidence of prostate cancer, meaning that, ultimately, routine screening with a PSA blood test is not advisable. During our informed-consent session, we talk about this very issue, namely, that if you want to have a PSA blood test, you should understand the inherent risks you are assuming.

## Testosterone replacement therapy does not increase the risk of prostate cancer!

I don't want to spend a huge amount time answering this question because it is not even close to controversial—it was answered years ago. For some reason, though, some healthcare practitioners still believe this myth. But that's all it is: a myth. **Testosterone replacement does not increase the risk of prostate cancer. Period.**

Numerous studies following men for long periods of time who were taking testosterone replacement therapy consistently show no alteration in the risk of prostate cancer. Meta-analysis studies have looked at all of the various studies, and there is no remaining controversy. You might still hear about a risk from physicians who aren't involved in the testosterone replacement field, but, otherwise, the fact that this "risk" is not a risk is very well established. (We will talk a little bit later about using testosterone in men with prostate cancer, which does have some controversy.)

## Let's talk about urinary tract symptoms

As men age, they start having symptoms of prostate enlargement. The prostate gets larger as men age, and it presses on the urethra, which is the tube that drains the bladder.

BPH (or benign prostatic hypertrophy) occurs in all men as we age. It is not cancer, and it may not be symptomatic, but the prostate enlarges as we age. When we are in our teens, it is very tiny—about the size of a tree nut. As we age, it may become the size of it peach. In some men, it might get much larger than that, which can give them urinary symptoms such as difficulty in voiding, emptying the bladder, and initiating urination. There are medications to treat this, and there are some supplements men take to avoid this, but it is a fact of life—as you age, your prostate will get bigger. It's true that some evidence has shown that when men initiate testosterone replacement therapy, their prostate volume increases slightly. What is not clear is whether such a slight increase would have happened without testosterone replacement. Another question is whether that slight increase has any clinical significance.

A study published in *World Journal of Men's Health* in 2013 by the Korean University School of Medicine evaluated the impact of testosterone replacement therapy on patients with lower urinary tract symptoms who were not taking medicines for BPH. In this study, they had a questionnaire called the International Prostate Symptom Score (IPSS). Men filled it out to provide a measurement of prostate symptoms, which are mostly related to urination. In this study, researchers included several hundred men taking long-term testosterone replacement who had urinary symptoms and BPH but were not on any medications for it yet. The men's symptoms substantially improved over the course of the testosterone replacement treatment. In short, testosterone replacement improved their prostate symptom scores.

A study out of Germany that appeared in the *World Journal of Urology* in 2000 followed 14,269 men with low testosterone who had symptoms of erectile dysfunction and had lower urinary tract symptoms such as difficulty voiding, inability to empty the bladder, and urgency. They put the men on testosterone replacement and followed them for five years. (They also used the same IPSS questionnaire.)

The researchers in this long-term study found that testosterone replacement therapy is not associated with worsening urinary tract symptoms. In fact, they found that over time, men develop fewer urinary tract symptoms the longer they are on testosterone replacement. There's a direct correlation between long-term use of testosterone and a *lower* incidence of urinary symptoms, not more.

## Testosterone replacement after prostate cancer

As you can imagine, this topic is controversial, so we're going to go over trends in the literature. A study published in the May/June 2016 issue of the *Canadian Urological Association Journal* surveyed Canadian urologists regarding their opinions about prescribing TRT in the wake of prostate cancer and how often they actually did prescribe testosterone replacement therapy for men who had had prostate cancer previously and were being monitored for future cancer.

In the survey, 63% of the urologists said that they believed testosterone replacement therapy does *not* increase the risk of progression of prostate cancer in men who have had prostate cancer and have been treated for it. Still, the urologists' words do not translate into actions when it comes to patient care. Of those surveyed, 65% of them stated

they would offer testosterone replacement therapy to men with low testosterone who have inactive prostate cancer . . . but only 42% of them
had ever actually prescribed testosterone to men who had previously
had prostate cancer and were actively being monitored.

A paper published in the *International Society for Sexual Medicine*
in 2013 conducted a literature search on testosterone replacement therapy in men with prostate cancer to determine what the direction of the
prevailing evidence was—after all, prostate cancer is very common. It's
the second most common cancer in men after skin cancer. (Men typically don't die of skin cancer, either.)

There is some evidence that the incidence of prostate cancer is higher
in men who have low testosterone—specifically, two studies have shown
that men with low testosterone have more incidents of prostate cancer.
One of those studies showed that men with low testosterone have a prostate cancer risk of 38.9%; in comparison, men with normal testosterone
have a prostate cancer risk of 29.5%. The other study showed that the
lower the testosterone level drops, the higher the risk of prostate cancer
a man has. However, this cannot be translated into thinking that testosterone replacement reduces prostate cancer risk. There is some evidence
that it does, but that evidence is not sufficient to draw a conclusion.

Another literature search looked at 46 studies conducted on testosterone replacement after prostate cancer. The researchers determined
that the studies collectively suggest that, assuming the cancer is no longer active, testosterone replacement therapy is safe in men who have had
prostate cancer. The researchers acknowledged that long-term studies
are needed before this can be conclusively stated but that, based on current data, they feel that it's safe to prescribe testosterone replacement
therapy for men who had previously had prostate cancer.

Still, there is insufficient evidence to definitively state that testosterone replacement therapy after prostate cancer is safe. We do know from
large studies that testosterone replacement therapy in men who have low
testosterone reduces their chance of all-cause death in the subsequent five

years by 50%. We do have evidence that testosterone replacement therapy can safely be given to men after treatment of prostate cancer. As of right now, however, we cannot say that testosterone replacement therapy for men who have had prostate cancer will extend their life the way it does for men who take TRT and have never had prostate cancer. It probably does; it certainly makes sense that it would. But we cannot say for sure.

Currently, the most practical approach to decide whether to treat a man with prostate cancer with TRT is to first assess how aggressive his cancer is and how likely it is to recur. If the cancer is relatively nonaggressive, move to step two, which is to determine the kind of impact testosterone replacement will have on the man's life. If he has cardiovascular disease, is overweight, suffers severely from other health issues, and also has low testosterone, testosterone replacement therapy will likely substantially improve his quality of life as well as his duration of life. If a man has fairly aggressive cancer that has a higher chance of recurring and the man is otherwise perfectly healthy and in shape with a borderline testosterone level, then it would make sense to withhold testosterone replacement therapy.

## Testosterone replacement and fertility

Testosterone levels in men start to decrease in very small amounts every year starting in their mid- to late 30s. Over 10% of testosterone replacement prescriptions are prescribed to men under 40 years old. These men presumably may still want to have children, so for them, at least, fertility is certainly a concern.

When a man takes testosterone (either by injection, pellet, or cream), the testosterone signals the pituitary gland in the brain to decrease the secretion of FSH and LH. FSH and LH are hormones the brain synthesizes to signal the testicles to make sperm and testosterone. Since testosterone is being taken at an adequate level, the brain signals the testicles to take a break from producing FSH and LH. Long-term suppression of FSH and LH can lead to a decrease in the formation of sperm.

There are many misconceptions regarding testosterone replacement and fertility. One study shows that 25% of urologists—those are physicians who specialize in treating infertility issues—mistakenly believe that testosterone replacement will improve fertility. This is not correct. However, there is also some evidence that analogs to FSH and LH (as well as other hormone-stimulating drugs) can restore fertility. And they do, to some degree.

Let's look at the latest evidence. A study published in *Fertility and Sterility* in 2016 looked at fertility and long-term testosterone replacement. The study emphasized that the "age and duration of testosterone therapy predict time to return of sperm count after human chorionic gonadotropin therapy." In this study, researchers evaluated men who experienced infertility as a result of long-term testosterone replacement by conducting lab tests such as evaluating sperm count and hormone levels. They gave the men 3,000 IU of hCG, a medication designed to stimulate the testicles to produce testosterone and sperm, and also a medication designed to upregulate the testicles.

What the researchers found was that about 70% of men *did* achieve resumption of satisfactory sperm counts with this treatment; however, the older the men were, the lower the odds were of the treatment working. Also, the longer the men took testosterone, the less likely the treatment was to work.

Certainly, 70% is better than 0%, but 70% is nowhere near 100%. When considering this treatment, then, you should bear in mind that the longer you take TRT and the older you are, the less likely you are to be fertile. This is a significant consideration if a younger man has low testosterone and plans to have a family.

## In summary

There are many misunderstandings regarding testosterone replacement therapy and urinary tract health. We've talked about the PSA blood test and how it is not beneficial—in fact, it's more likely to be harmful than helpful—but that in spite of this, many medical societies still recommend doing a PSA blood test. If a man chooses to do a PSA blood test when he starts testosterone replacement therapy, he should do so with informed consent, knowing that the test is not beneficial. Since testosterone replacement does not increase the risk of prostate cancer, initiating testosterone replacement therapy does not alter the current guidelines that state that the use of the PSA blood test should be discontinued.

There are many misconceptions and myths regarding testosterone replacement therapy and the development of prostate symptoms and prostate cancer, but the truth is that testosterone replacement therapy does *not* increase the risk of developing prostate cancer. In fact, testosterone replacement therapy potentially could protect *against* prostate cancer. However, adequate proof of that does yet not exist.

Testosterone replacement therapy does improve symptoms associated with enlarged prostate, such as urinary frequency, urgency, and hesitancy. Men who have prostate symptoms do not need to avoid testosterone replacement therapy.

However, testosterone replacement therapy—particularly if taken for a long period—will reduce fertility. There may be ways to offset this that have yet to be discovered, but this issue remains a real concern. If you have low testosterone and are considering testosterone replacement therapy, you need to weigh the benefits of testosterone replacement therapy against the risk of becoming infertile.

# 8

# Testosterone and Your Musculoskeletal System

Let's talk about testosterone replacement therapy with respect to bones, muscle, and lean body mass. As men age, their testosterone levels generally decline, which leads to a decline in bone density and density of fat-free mass. In the face of low testosterone, muscle size can diminish in spite of exercise.

There are two types of testosterone replacement:

1. Anabolic steroids (used for bodybuilding)

2. Treating low testosterone

## Anabolic steroids for bodybuilding

Let's define the differences that exist between using testosterone replacement for low-testosterone syndrome and using anabolic steroids to become unnaturally large (the latter is what bodybuilders may do). Evidence shows that if a man has a normal testosterone level and he takes the kind of supraphysiologic and synthetic testosterone doses that bodybuilders take, he will develop larger muscles when he exercises. To achieve larger-than-normal muscle gains requires taking larger-than-normal doses of testosterone and its synthetic derivatives.

Generally, bodybuilders will take synthetic testosterone analogs in order to achieve their supernatural physiques. To get the kind of unnatural (even unhuman) muscles that bodybuilders typically have, synthetic drugs are much more effective than massive doses of testosterone, so that's what they usually use.

Nandrolone, decanoate, and anabolic steroids are analogs of testosterone that are custom designed by chemists in order to have an anabolic effect (a.k.a. growth effect) on muscle fibers. Testosterone is necessary for muscle fiber growth, and these custom-made drugs manipulate the same receptors that testosterone affixes to, thereby causing muscles to overgrow, especially in response to physical training.

This has led to significant bias that all testosterone replacement is for bodybuilding. Yes, some bodybuilders abuse testosterone in order to grow muscle. In general, though, they use other synthetic analogs that work far better than testosterone—testosterone is a hormone, and the synthetic analogs are drugs that are more or less "super testosterone" drugs. When these drugs are taken with the intent of gaining muscle for bodybuilding purposes, they carry substantial risks of side effects, are not safe, and are not at all the topic of this book. Unfortunately, the use of these steroids—and the popular culture of bodybuilding that goes along with it—has created an association between anabolic steroids and replacing low testosterone with TRT. They are not at all the same. Testosterone replacement refers to treating a man who has poor health or diminishing quality of life as a result of having low testosterone with the intent of restoring his testosterone to a normal level.

## Testosterone replacement for low testosterone

This refers to testosterone supplementation (TRT) in men who have symptoms of low testosterone. Symptoms of low testosterone can vary, but they commonly include decreased energy and vigor, decreased libido, minor mood disturbance, weight gain, and fatigue. Low testosterone can also be associated with more serious conditions like diabetes, cardiovascular disease, congestive heart failure, hypertension, and obesity.

A study done in *Annals of Agricultural and Environmental Medicine* in 2014 evaluated the effects of testosterone replacement in men who were 45–60 years old in response to exercise regardless of their testosterone level. The researchers gave testosterone replacement in the form of a weekly 100 mg shot and had the men do various exercise protocols.

The men were split up into two groups: those receiving testosterone replacement and those receiving no testosterone replacement. All of the men weighed about the same, were in the same age range, and had the same amount of body fat. They were followed for 12 weeks. Their exercise protocol included exercising four days a week, with two days of strength training and two days of aerobic endurance exercise.

The men who were in the control group and who were given a placebo had different results than the men who were given TRT: the placebo men lost about one pound of fat, while the men taking testosterone replacement lost about six pounds of fat. The men taking TRT gained about four pounds of muscle mass, while the men on a placebo did not gain any muscle mass in spite of regularly exercising four days a week.

Many middle-aged men find that even though they diet and exercise, they don't lose fat like they want to, plus they feel like they're not gaining any muscle, either. This study demonstrated that testosterone replacement resolves this problem. Now, I'm not suggesting that if we put all men on testosterone, they will gain muscle mass and lose fat in response to exercise—some middle-aged men have normal testosterone levels and don't need replacement. In reality, though, most men's testosterone drops as they age. This is considered "natural." However, this drop also leads to natural consequences in some men, which include weight gain, obesity, diabetes, cardiovascular disease, and premature death.

The reality is that there are far more men with low testosterone than we would think. Testosterone declines every year after age 30, so most men in their 40s and 50s have significantly lower testosterone levels compared to when they were younger. This goes hand in hand with the diseases that occur as we age, diseases that are directly related to testosterone deficiency. The notion that we should treat only men with severely low testosterone caused by specific genetic conditions or organ damage is nonsense.

## What about older men?

A study published in the *American Journal of Physiology and Endocrinology Metabolism* in February 2014 conducted a randomized controlled trial. A randomized controlled trial means that one group of people are assigned to the experimental group and one group of people are assigned to the placebo group. Neither the investigators nor the individuals know which treatment they are getting (i.e., either active treatment or a placebo). This is a gold standard when it comes to conducting medical research.

This study evaluated men over 60 years old who had a testosterone level of 300 ng/dL or lower and their response to testosterone replacement or a placebo over one year. The men did not do any particular exercise or other interventions during that time. At the end of the year, when the researchers assessed the outcomes, the men with low testosterone and who had received testosterone replacement had gained about nine pounds of bone and muscle mass, experienced a 8%–14% increase in muscle strength, and lost about eight pounds of fat. As seen in other studies, there were no adverse effects. Men with moderately low testosterone and with a continued normal lifestyle had gained muscle mass and bone density and decreased their fat. Testosterone deficiency leads to more fat and less bone and muscle mass, and testosterone replacement reverses that effect.

A meta-analysis is a study that identifies multiple studies that were focused on similar issues by matching similar limiting definitions and parameters. Once those similar studies have been identified, a meta-analysis then compiles data from those multiple studies and tries to come to a logical conclusion based on the aggregated data so that doctors or scientists or any group of people looking at the studies can also draw their conclusions.

Numerous meta-analysis studies have reviewed what happens when men are given testosterone replacement as they age and naturally develop low testosterone. As I write this book, I'm looking at three journal articles in front of me that each describe thousands of men and hundreds of clinical trials showing that testosterone replacement in men with low testosterone improves body composition and lean muscle mass, decreases free fat mass, and improves metabolic profiles in men. No studies suggest otherwise. At the same time, studies show that low testosterone leads to decreased bone and muscle mass, while restoring testosterone reverses that. Period.

## Loss of bone density

The evidence is pretty clear: men with low testosterone have lower bone mineral density, which leads to more fractures, increased frailty, and more joint problems. Testosterone replacement improves bone mineral density in men with testosterone deficiency.

Studies have been done to assess bone mineral density in men with low testosterone who then go on to take testosterone replacement therapy. Bone mineral density is fairly easy to assess—various radiologic procedures can be done to accurately measure the density of bones. Regardless of the cause, testosterone deficiency leads to decreases in bone mineral density, and testosterone replacement leads to restoration of bone mineral density.

A study done in Spain and published in *Andrology* in 2013 evaluated the outcomes of men 50–65 years old with low testosterone and their response to testosterone replacement therapy. Researchers followed the men for 12 months and found a sustained, definitive, and consistent improvement of bone mineral density that began early on—right at the outset of treatment—and continued until the end.

Another study done in the *European Journal of Endocrinology* in 2005 evaluated men with an average age of 46 who had low testosterone and their response to testosterone pellets. Researchers followed the men for about three years and found that the men who saw a normalization of their testosterone levels to an ideal level showed the greatest improvement in bone mineral density, while the men with suboptimal replacement testosterone saw less benefit. Still, all of the men showed improvements in bone mineral density.

## In summary

In summary, low testosterone is associated with an increased loss of bone density, an inability to gain muscle mass, and an inability to lose fat even when exercising. Low testosterone is also associated with other quality-of-life issues. Testosterone replacement therapy in men with low testosterone is associated with an improved ability to gain muscle mass and bone density and a substantially improved ability to lose fat. Even in the absence of exercise, men with testosterone replacement tend to lose fat and gain muscle. That said, men shouldn't rely on hormone replacement as a means for maintaining health—it's best to use both TRT *and* lifestyle modification to maintain and improve health.

# 9

# Testosterone and Sexual Function

I t's common knowledge that low testosterone levels are associated with decreased sexual function and increased erectile dysfunction, and yet there are some who claim that isn't the case, namely, FDA policymakers—they say that decreased sexual function and erectile dysfunction associated with normally declining testosterone levels are not indicators for testosterone replacement.

In early January of 2017, at the British Fertility Society conference in Edinburgh, Professor Huhtaniemi attacked the use of testosterone for treating low testosterone, saying that conditions that are clearly identified in scientific literature as being related to low testosterone are in fact just a result of getting older. The professor claimed that testosterone replacement has little impact on quality of life and is not worth its "risks." As he put it, "I tell people that taking testosterone will give you a strong handshake, but really not very much more than that."

He is just one of the many people who continue to cite the flawed and since-retracted study that erroneously said that TRT should be avoided due to an increased risk of heart disease or stroke, a statement that has been soundly disproved and rejected. The full article is available here: www.dailymail.co.uk/health/article-4100584/Warning-testosterone-jabs-help-men-male-menopause.html.

## Adapting early to new science

When things change in health care, when we find out about something new that we didn't previously know about (or believe), or when we find out that our assumptions and guesses have been proven to be just plain wrong, doctors should be early adapters to new science, because an overwhelming wave of new information can change the practice of medicine. This way of thinking is called practicing "evidence-based medicine." It is what we teach doctors during their training: "What I am teaching you about today will change, so you must understand that as a physician, you need to constantly be learning. The way you practice medicine five years from now will be different from the way I'm teaching it to you now."

However, some physicians cannot accept this. They feel they were trained by experts and that they are experts, so when they see information that conflicts with their expertise, they choose to ignore it the way the aforementioned professor did. Generally, information is selectively ignored, as evinced by that professor who ignored the overwhelming evidence when he said that there is no benefit from testosterone replacement therapy. (It's worth noting that during the same conference, he chose to selectively remember that he thinks testosterone replacement causes heart disease.)

My instinct is to learn, interpret what I learn, and adapt. Some physicians avoid learning in order to avoid adapting; some physicians learn, choose to ignore what they learn, and don't adapt. There is a quote that sticks in my head when I think about the changes that have occurred since I started practicing medicine. Back in 2004, I heard this quote repeated during a lecture I attended. I have no idea where it originated, but the quote goes like this: "For every pioneer in medicine, there are 1,000 self-appointed guardians of the past."

## Let's look at the science

In March of 2015, the *Journal of Clinical Endocrinology and Metabolism* published a study evaluating the sexual function, vitality, and physical functioning of older men with low testosterone levels. The study evaluated men over 65 years of age with a testosterone level lower than 275 ng/dL (which is very low). Several hundred men participated in this study. The researchers found a statistically significant and consistent correlation between lower testosterone levels and lowered sexual desire, erectile function, and sexual frequency. Remember that the FDA specifically states that testosterone replacement therapy should not be used because of decreasing testosterone associated with "normal aging." This contradicts logic.

Even more striking than the linear relationship between low testosterone and low sexual function in older men is the association between low testosterone and lower sexual function in younger men. Low testosterone in younger men has a more profound effect on sexuality, regardless of the cause.

When we look at symptoms associated with low testosterone, sexual dysfunction is one of the most specific complaints men have and why they may seek out medical attention. While there may be a variety of reasons for men to have low sexual function, erectile dysfunction, and/or decreased sexual frequency, low testosterone is a fairly consistent reason. Probably the only thing that is more consistently associated with low testosterone is obesity or belly fat.

Another study published in 2016 in the *Journal of Urology* evaluated men who presented for urologic evaluation of erectile dysfunction and had their testosterone levels evaluated. Although the researchers found various reasons for erectile dysfunction (including obesity and advanced age), low testosterone levels were consistently associated with sexual dysfunction.

Numerous studies have shown a link between low testosterone and sexual dysfunction, which can include erectile dysfunction, decreased sexual arousal in terms of libido and desire, and decreased sexual frequency. The reason low testosterone leads to sexual dysfunction and erectile dysfunction is becoming clearer as science continues to progress—for example, the class of drugs called PDE5 inhibitors (think Viagra) focused a significant amount of research on men's sexual health and brought such concerns to the forefront of public awareness. There does appear to be a direct relationship between PDE5 and testosterone and other androgens, and testosterone does have a direct effect on the penile tissue. It has long been known and observed that testosterone replacement improves libido, erectile function, and sexual frequency.

In 2004, in a study published in *Reviews of Urology*, researchers measured the response time that men with low testosterone who took testosterone replacement experienced in terms of their sexual health. Questionnaires used to assess their sexuality were given to the 638 men in the trial who were given testosterone replacement for 30 days. On the questionnaire, the men were asked to rate their responses to the following three questions, with 0 being none and 7 being very high:

1.  Please rate your overall level of sexual desire today on a scale of 0 to 7.

2.  Please rate the level of enjoyment or pleasure of any sexual activity you experienced today without a partner (e.g., masturbation, sexual fantasies) on a scale of 0 to 7.

3.  Please rate the level of enjoyment or pleasure of any sexual activity you have experienced today with a partner (e.g., kissing, intercourse) on a scale of 0 to 7.

Then more specific questions followed:

4.  Please indicate if a partner is available, yes or no.

5.  Please rate your mood for each, from 0 to 7, where 0 indicates not at all and 7 indicates very true: angry, alert, irritable, full of pep/energetic, sad or blue, tired, friendly, nervous, well/good.

6.  Circle any of the following that you have experienced today:

Sexual daydreams, anticipation of sex, sexual interactions with your partner, flirting (by you), orgasm, flirting (by others toward you), ejaculation, intercourse, masturbation, nighttime spontaneous erection, daytime spontaneous erection, erections in response to sexual activity.

7.  If you experienced an erection today, indicate the percent of full erection you experienced.

8.  If you experienced an erection today, on a scale from 0 to 7, indicate whether it was maintained for a satisfactory duration.

The average age of the men in this study was 53, with an overall age range from 18 to 86. The men had an average morning serum testosterone level of 204 ng/dL, which is very low. They began using testosterone after they had completed the questionnaire.

Some men experienced improvements (relative to most of the questionnaire questions) beginning as early as seven days after starting TRT and continuing throughout the duration of the one-month study. For example, prior to treatment, the men were having sexual intercourse less often than one time per week. Toward the end of the month-long study, their sexual intercourse averaged more than one time per week. The treatment didn't greatly change the men's lives, but it did help them be more sexually active and more sexually interested, plus it improved their sexual dysfunction.

Another study in 2009 in *Endocrinology and Metabolism* evaluated the response to testosterone replacement in men with low testosterone, comparing men who took TRT to men who relied on vitamin replacements. The researchers used the IIEF-5 (International Index of Erectile

Function) questionnaire—which you will have a chance to answer later in this chapter—and the ADAM questionnaire (Androgen Deficiency in Aging Males).

## The IIEF-5 Questionnaire

1.  How do you rate your confidence that you could keep an erection?

| 1 | 2 | 3 | 4 | 5 |
|---|---|---|---|---|
| Very low | Low | Moderate | High | Very High |

2.  When you had erections with sexual stimulation, how often were your erections hard enough for penetration (entering your partner)?

| 1 | 2 | 3 | 4 | 5 |
|---|---|---|---|---|
| Almost never or never | A few times (much less than half the time) | Sometimes (about half the time) | More times (much more than half the time) | Almost always or always |

3.  During sexual intercourse, how often were you able to maintain your erection after you had penetrated (entered) your partner?

| 1 | 2 | 3 | 4 | 5 |
|---|---|---|---|---|
| Almost never or never | A few times (much less than half the time) | Sometimes (about half the time) | More times (much more than half the time) | Almost always or always |

4.  During sexual intercourse, how difficult was it to maintain your erection to completion of intercourse?

| 1 | 2 | 3 | 4 | 5 |
|---|---|---|---|---|
| Extremely difficult | Very difficult | Difficult | Slightly difficult | Not difficult |

5.  When you attempted sexual intercourse, how often was it satisfactory for you?

| 1 | 2 | 3 | 4 | 5 |
|---|---|---|---|---|
| Almost never or never half the time) | A few times (much less than the time) | Sometimes (about half half the time) | More times (much more than | Almost always or always |

## The ADAM questionnaire

Table 1. Questions Used as Part of the Saint Louis University ADAM Questionnaire

1.  Do you have a decrease in libido (sex drive)?

2.  Do you have a lack of energy?

3.  Do you have a decrease in strength and/or endurance?

4.  Have you lost height?

5.  Have you noticed a decreased "enjoyment of life"?

6.  Are you sad and/or grumpy?

7.  Are your erections less strong?

8.  Have you noted a recent deterioration in your ability to play sports?

9.  Are you falling asleep after dinner?

10. Has there been a recent deterioration in your work performance?

NOTE. A positive questionnaire result is defined as a "yes" answer to questions 1 or 7 or any 3 other questions.

The researchers followed these men for eight weeks. Based on responses to these surveys, the men who were taking testosterone saw an improvement in their libido and erectile function. However, as expected, the men with low testosterone who took the vitamins (as a placebo) saw no change.

A study reported in the *Journal of Sexual Medicine* in 2016 examined a randomized double-blind, placebo-controlled, 16-week-long study of 750 men with low testosterone who also had a low sex drive. The men were given either a placebo or testosterone replacement. The researchers found that for the most part, the men taking testosterone consistently experienced a higher sex drive, improved erectile function, and increased energy. The lower their testosterone level was at baseline, the more significant their improvements were in the follow-up portion of the study.

A study published in the *New England Journal of Medicine* on November 29, 2016, evaluated 790 men ages 65 and older with low testosterone levels. The men were put on either a placebo or testosterone for one year. In addition to sexual function, the study also evaluated other things such as mood and depression (which, like sexual function, improved). Researchers also noticed that the higher the men's testosterone level rose during therapy, the more improvement they saw in their sexual function: their sexual desire increased, their erectile function improved, and they had an increase in sexual frequency. Again, this happened in older men, whose sexual function concern is often ignored.

A 2016 study done in the *British Journal of Urology International* evaluated men with low testosterone and type 2 diabetes who took TRT for more than 30 weeks and compared their results to men with low testosterone and type 2 diabetes who took a placebo for 30 weeks. The researchers observed improvements in sexual symptoms and function in the men taking TRT. These men, it should be noted, may have had more than one reason for low sexual function and erectile dysfunction—both low testosterone and diabetes can cause decreased sexual function. The men's sexual function (as measured by sexual desire) improved as early as six weeks into the study. Erectile dysfunction began to improve toward the end of the study, suggesting that longer-term use of testosterone therapy may be needed in men with low testosterone and erectile dysfunction who are also diabetic.

## In summary

Low testosterone leads to lower sexual function, but testosterone replacement can restore that sexual function or at least improve it. Low testosterone is also related to erectile dysfunction: the lower the testosterone is, the more erectile dysfunction there is. Again, testosterone replacement can improve that. When we look at how many prescriptions are written for Cialis, Levitra, and Viagra every year, we have to ask: how many of these men are checked for low testosterone?

There is no doubt that men have improved sexual function when their low testosterone levels are restored to a normal level. These improvements include an increase in sexual fantasies, sexual desire, and sexual sensation and improved erectile function. Studies show that the lower a man's testosterone level is, the more significant his sexual dysfunction will be. Raising a man's testosterone level up higher—that is, closer to the level a normal young man would have—will have the most benefits in terms of restoring sexual and erectile function.

Additionally, common parameters like age, weight, or medical history should not determine whether testosterone therapy should be used for treating sexual dysfunction: it works well in older men, it works well in obese men, it works well in men with diabetes. These are common conditions that sometimes prompt physicians to ignore a man's concerns regarding his sexual health, yet sexual health remains an important concern for men. While you can't do anything about your age, you *can* do something about low testosterone, plus TRT has beneficial effects on obesity, metabolic syndrome, diabetes, and sexual and erectile function.

According to *Harvard Health Publications* (published by the Harvard Medical School), about half of men ages 40–70 have erectile dysfunction to at least some degree and about 10% are completely unable to have erections. This reality has led to a huge increase in the sales of erectile dysfunction drugs . . . yet testosterone levels are infrequently checked in these men. See the full write-up at www.health.harvard.edu/mens-health/which-drug-for-erectile-dysfunction.

Let's close this chapter with some health concerns that affect men and that are directly correlated with low testosterone:

- One in four men will eventually die from heart disease. One of the largest studies ever conducted shows that men with low testosterone are about half as likely to die prematurely from heart disease if their testosterone is replaced versus if their testosterone deficiency is not treated.

- Well over 50% of men are overweight or obese. Testosterone deficiency is directly correlated with weight gain, and testosterone replacement results in sustained long-term weight loss and weight maintenance.

- Erectile dysfunction and decreased sexuality are associated with testosterone deficiency, and testosterone replacement has been shown to improve these conditions as well. But how often is a man's testosterone level checked?

# 10

# Testosterone and Mood/Brain

The beneficial effects of testosterone on mood, vigor, vitality, and energy are fairly well established and have been noted in the numerous clinical trials we've already discussed, but let's talk more about these mood- and brain related aspects of TRT. There are some questionable areas of benefit with respect to testosterone replacement, too, which we'll address. Some of these areas are still being studied and may offer promise in the future.

## Does testosterone enhance mood?

In *Archive of Urology* in September of 2013, there was a discussion about the increasing prevalence of depression in men as they age, which mirrors the increase in the prevalence of low testosterone. The researchers noted that testosterone replacement improves depressive symptoms in most men with depressive symptoms. Studies also demonstrate the chemical interactions that exist between depressive chemicals and testosterone.

Some experts recommend that men who are being evaluated for depression be given a testosterone replacement therapy trial to assess their response even if they're already taking medication for depression. Again, we know that testosterone levels—or, at least, absolute testosterone levels—do not correlate well with symptoms of low testosterone. Perhaps just giving TRT a try to see whether it can alter something as quality of life-

A study published in the February 2016 issue of the *New England Journal of Medicine,* "Effects of Testosterone Treatment in Older Men," evaluated the response of men aged 65 and older who took testosterone with respect to their moods and depressive symptoms. This study involved 790 men with low testosterone. About half were given testosterone, while the other half were given a placebo. There was a direct and significant improvement in the mood of the men who were on testosterone replacement; their depressive scores also improved. The men on a placebo did not see any such changes.

A study published in *Clinical Interventions in Aging* in 2015 (conducted at the Medical University of Warsaw in Poland) evaluated men with prediabetes with respect to their depressive symptoms. Researchers also studied various laboratory parameters, including the men's testosterone levels. They found a strong correlation between low testosterone and depressive symptoms and recommended that testosterone levels be evaluated in men with depressive symptoms as well as men with prediabetes.

These observations and studies evaluating the relationship between testosterone and depression are not new—I've just been presenting the more recent ones. This next study I'm going to talk about goes back to August of 1992, however, and it's from the *Journal of Andrology*. It was conducted before we understood how much impact low testosterone has in men. The researchers evaluated six men who had profoundly low testosterone due to developmental issues, including a lack of ability to develop normal testicles.

The researchers did some testing on their sexuality, such as checking for nighttime erections (the men had none) as well as measuring other aspects of their sexuality. In addition, the researchers also measured the men's level of depression. Then the men were put on testosterone replacement therapy; specifically, the researchers elevated their levels to reflect the levels normal healthy men would have. The result? Their depression scores substantially improved, as did their sexuality.

And these connections are not just being studied in America and Europe—they've been studied in Asia, too. A study published in 2012 in the *Chinese Medical Journal* randomized 160 men with low testosterone into two groups, one group of men who received testosterone replacement and one group who received a placebo. The researchers measured their stress scores, anxiety scores, and depression scores and found that testosterone replacement improved not only the men's low testosterone symptoms but also all aspects of their psychosocial life (i.e., TRT lowered their anxiety, stress, and depression).

## There are different types of depression

"Atypical depression" occurs when someone becomes depressed but gets better in response to positive stimuli. An example of atypical depression would be a man who finds himself depressed with some frequency, but then, when he is exposed to something very positive (such as his children visiting), his depression goes away or gets better and his mood improves.

Then there is typical depression, also known as "melancholic depression," which is when someone's mood does not improve in response to events that would normally be pleasurable.

Many studies have looked at men's response to testosterone replacement therapy in terms of depression. Generally, men have a positive response to treatment, but still, not all men respond. Why not? To answer this question, psychologists and psychiatrists from Germany and Brazil evaluated the testosterone levels of men with either typical (melancholic) depression or atypical depression.

They found that the men with melancholic depression tend to have low testosterone levels to some degree. Men with atypical depression, however, have *much* lower testosterone levels than those with melancholic depression. This difference in subtype might be why studies done on men with general depression show inconsistent results (although men generally have a positive response to TRT). This may suggest that men with melancholic depression have more of a brain chemical imbalance, while men with atypical depression may have more of an "irritable man" kind of syndrome, or what I like to call "grumpy old man syndrome."

## Memory and cognition

There's a lot of interest in determining whether testosterone replacement can preserve cognitive function such as memory, language skills, and thought processes. Studies conducted on animals have shown that testosterone replacement does reduce cognitive decline by reducing oxidative stress on the brain.

Studies in humans, however, have failed to consistently show any evidence that maintaining healthy testosterone levels (either naturally or with testosterone replacement) prevents cognitive decline, something that naturally occurs with aging. People may believe that TRT maintains cognition, but, unfortunately, there isn't good evidence to support it. That said, testosterone replacement does not *worsen* cognitive decline. In short, other than noting that a few studies have shown some minimal benefits, we should not assume that testosterone replacement will preserve memory or cognitive function.

A study published in *World Journal of Men's Health* sought to address this issue. Researchers identified 106 men with an average age of 57 who had low testosterone levels and were symptomatic. One group was given testosterone replacement therapy; the other was given a placebo.

At the end of eight months, the men who were given a placebo had no change in their serum testosterone levels, their erectile function, or their depressive symptoms. They also had no improvement in their cognitive functioning.

In contrast, the men on testosterone replacement therapy saw improvements in their testosterone levels, erectile function, cognitive function, and depression scores. What's particularly interesting to me is that the researchers took men with very low testosterone levels and raised their testosterone to being *merely* low instead of *very* low. Because conventional wisdom dictates that a "normal" testosterone level is based on laboratory results alone, we still don't know how much testosterone replacement therapy can potentially improve cognition or prevent cognitive decline.

## Parkinson's disease and other brain disorders

A few studies have evaluated testosterone therapy in men with Parkinson's disease. Again, these results have been positive, but they weren't consistent. The studies have been neither large-scale nor long-term, either, but they have shown promise that TRT may benefit non–motor symptoms of Parkinson's disease. Men with Parkinson's disease may have low testosterone just like any other man might, and these men may likewise benefit from testosterone replacement therapy if they have low testosterone. However, providing testosterone replacement therapy for the sole purpose of improving Parkinson's disease is not justified.

Testosterone is also being studied in conjunction with Alzheimer's disease, but again, the results have not been at all consistent. While there is some evidence that memory and certain very selective functions improve with testosterone replacement in patients with Alzheimer's disease, this is by no means definitive. Testosterone should not be used as a primary treatment for Alzheimer's disease. That said, if men with Alzheimer's disease have other reasons for needing testosterone replacement therapy, then certainly they can receive treatment.

There's also some interesting ongoing research regarding testosterone replacement therapy and neurological diseases such as multiple sclerosis. While this research is still in an experimental phase, so far, testosterone does appear to be beneficial for treating men who are starting to have symptoms of multiple sclerosis. Again, though, we are not at a point where TRT can be universally recommended for this condition because (so far) we lack sufficient evidence.

## In summary

At this point, there is significant evidence that testosterone replacement therapy improves men's mood and depressive symptoms if they also have low testosterone. Because the subtype of men with atypical depression (meaning their depression gets better when they're exposed to positive events) will likely experience more significant benefits with testosterone replacement therapy, the use of TRT should be considered in this setting.

Testosterone replacement therapy for slowing down cognitive decline in brain function and memory preservation is not likely to yield any benefit, although it will not hurt. Testosterone replacement therapy should *not* be used for the sole purpose of preserving memory and brain functions.

How testosterone replacement therapy affects brain conditions has been evaluated but has yielded inconsistent results. While TRT is not harmful for men with neurological conditions, testosterone is not likely to cause any primary significant benefit for these conditions. Testosterone replacement therapy should be considered for men with low testosterone who have symptoms of low testosterone, but TRT is *not* a primary treatment for neurological diseases.

Testosterone replacement has the potential to benefit certain types of depression in that TRT certainly improves mood, vigor, and mental energy. Testosterone replacement in men with very low testosterone can potentially also improve their cognitive abilities (i.e., their ability to think). While men with various neurological diseases may not directly benefit from testosterone replacement—more studies need to be conducted on this topic—there's certainly no evidence that testosterone replacement therapy has a negative effect on the brain or mood, meaning that there is no reason to withhold TRT because a man is experiencing neurological problems.

# 11

# Treating Low Testosterone

There are several ways to initiate TRT. The very first options were by injections (shots into the muscle) and then pellets. Pellets are a slowly dissolvable form of testosterone inserted into the fat. A tiny hole is made after you are numbed up, a device is used to insert the pellet, and, because the incision is so small, it is taped shut and doesn't need to be stitched. There are newer forms on the market that include gels as well as a nasal spray.

Since testosterone has been FDA approved since the 1950s, drug manufacturers have devised different ways of delivering it, so they can get a patent and FDA approval and generate new business. New forms are showing up continuously.

So what is the best option? We are near the end of the book, so I won't present several more clinical studies, as the information here is taken largely from the studies I have already written about.

Longer-lasting forms are better. The studies with the most profound results had the longest-acting forms of testosterone. The forms that are the longest acting are testosterone pellets, which last about five months, and testosterone undecanoate, which is a long-acting shot that lasts about 10 weeks. Undecanoate was recently approved in the United States and has been available in Europe for years. Pellets are less expensive and last longer, so we use these primarily.

The next best options are standard testosterone injections. These were designed to be given every two to four weeks but are much better suited to be used every week, which leads to more of a steady state level.

Gels are the least desirable choice. Although very easy, they are applied twice a day and can be transferred to other people, and the profound results of TRT we have discussed in this book were more notable with injections or pellets.

Pellets are our current first choice.

That said, the reality is that any form of testosterone replacement for men with low testosterone is preferable to none, but know that testosterone pellets are currently the best option.

## What about health insurance coverage?

Coverage varies from insurance company to insurance company. They generally have some sort of policy that they don't necessarily enforce. For example, Blue Cross Blue Shield of Michigan, which is the most common insurance that my patients have, approves testosterone replacement in men with symptoms of low testosterone, which they define as loss of height or lack of sexual development (even though they don't cite the FDA's indications, lack of sexual development would typically be due to a chromosome abnormality or disease to the brain or testicles at a young age). Also, the testosterone level needs to be below "normal"—which Blue Cross defines as needing to be below 280 ng/dL—on two different blood tests taken in the morning. (The reason for requiring that testosterone be tested in the morning is that men's testosterone level is naturally at its highest in the morning.) Basically, only about 5% of the men who have low-testosterone symptoms and would benefit from testosterone replacement therapy are going to qualify under insurance guidelines. Additionally, those guidelines apply only if you're under 50—if you're over 50 years old, Blue Cross of Michigan requires that your total testosterone level be below 200 ng/dL. It's amazing the insurance company wants you to be really sick before it will pay for testosterone replacement that alleviates the diseases it *is* willing to pay for, such as diabetes and heart disease. This policy simply makes no sense, but that's the way it is. Also, in order to receive testosterone replacement, it requires that men not have any significant prostate symptoms . . . even though testosterone replacement has been proven to improve prostate symptoms.

Now, in reality, a doctor might prescribe testosterone replacement in a man who does not meet these criteria, and in this case, there is a chance the insurance company will pay for it. It is not until the doctor's records are audited or the patient's records are audited that the insurance company could ask for the money back.

But let's say that you actually *do* meet the strict criteria from Blue Cross Blue Shield of Michigan for it to pay for your testosterone replacement. If you qualify, it will pay for testosterone pellets as well as other forms. However, before it will pay for pellets, it requires you to have failed in the use of (cheaper) shots of testosterone first, and, what's more, it will pay only for 450 mg of testosterone rather than the usual dose of 2,000 mg. This is based on the concept that Blue Cross relies entirely on laboratory values, and it will pay only to get someone from a severely low level to the low end of normal levels.

When insurance does pay for cheaper forms of testosterone (like shots), it pays only for 50 mg given every two weeks. In our practice, we find that men require at least 100 mg every week to experience any benefits. As we've talked about throughout this book, there's also an old-fashioned concept that the blood test is the sole determinant for measuring low testosterone and that men have to have severely low testosterone to quality for treatment. Even when treatment is given, the logic goes, it should take a man only from a very low level to the low end of normal. This is in absolute contradiction to what the medical literature says. You can check with your own insurance carrier to see whether it is equally old-fashioned. It probably is.

This is not how I'm going to practice medicine, so we do not take insurance for the pellets—doing so would require me to compromise my ethics and undertreat men. Therefore, you have to pay for it yourself. It's frustrating when you pay your insurance premiums and the company then withholds covering your treatments, but that's the way insurance often works.

Are pellets FDA approved?

Yes, pellets are FDA approved. There are two types of pellets commonly used. One is Testopel, which is made by Endo Pharmaceuticals. We find men generally need about 10 mg per pound of body weight to get a level that has substantial improvements on quality of life. So a 200-pound man would require 2,000 mg every five months, and this would cost more than $6,000 annually using this product. We do not use it because of the cost.

The pellets we use currently are from AnazaoHealth, which is a smaller compounder of testosterone pellets, and the company is approved by the FDA as a 503(b) facility. There are several 503(b) FDA-approved facilities, so this has led to competition and has driven the cost down substantially, about a third of what the large drug companies' product costs.

## What will I feel like after I have testosterone replacement?

It takes about three weeks to notice the effects of pellet-delivered testosterone, and then you'll continue to notice improvements for about a year. We've talked a lot about the benefits of testosterone replacement and the fact that various improvements happen at various time intervals. Things such as sexual interest improve relatively quickly, for example, while things such as weight loss occur more slowly. (Weight loss typically occurs continuously in obese men until they reach a normal weight or close to a normal weight, but the loss doesn't happen right away—it increases a little bit every year.)

Symptoms like poor erectile function may take up to six months to significantly improve; after that, improvements are sustained. Energy, mood, and vigor typically improve gradually over several months. More significant aspects such as cardiac protection develop over years. For example, if a man's going to inevitably have a heart attack in a month because of his risky lifestyle factors and he starts taking testosterone, this will not prevent him from having a heart attack caused by smoking, weight gain, high cholesterol, or various other factors (in addition to low testosterone). However, if a man who is overweight and who has risk factors for heart disease, hypertension, and/or diabetes starts testosterone therapy, his risk factors will go down and his protection will go up.

In prior chapters, we talked about the effects of testosterone replacement on most organ systems. Long-term studies show that the longer the man is on testosterone replacement, the more significant the benefits are. What we see in clinical practice is that men feel better in a few weeks, mostly in terms of improved energy, sleep, vigor, and libido. Other health benefits gradually occur over much longer periods of time.

## The next step

If you're reading this book, either you are on testosterone replacement therapy and want to learn more about it or you may think you're suffering from low testosterone but need to get more information before considering replacement. Or perhaps you're reading this book out of general interest.

If you're just reading this book out of general interest, I hope you enjoyed it. This was a fun book for me to write! I like exploring controversies, and I love to learn, read, and teach.

If you're on testosterone replacement and you wanted to get more information, I also hope you enjoyed the book. We talked about the different forms of testosterone replacement, whether that's creams, gels, injectable testosterone, or testosterone in pellet form. If you are happy with what you're using now and you are compliant—that is, you are able to use it exactly as directed—there is no real reason to change what you're using if you feel you are getting adequate benefits. I would assume your healthcare provider is monitoring your hormone levels—we check things like estrogen and DHT to assess whether the testosterone is turning into a hormone that is maybe not so desirable. This situation is easy to manage.

No absolute blood level predicts low testosterone; likewise, no absolute blood level predicts adequate testosterone. That said, however, we find that most men feel normal with testosterone levels between 600 and 1,100 ng/dL.

If you're a man who is thinking you may have low testosterone yet your healthcare practitioner has not adequately addressed it, I suggest you seek another opinion. It is very common in the United States to markedly undertreat low testosterone. First, there is the bias of the blood test, and then—as we've discussed—there are the FDA's misguided rules that put fear into the minds of many physicians. These unfortunate guidelines will have untold public health consequences—it's estimated that hundreds of billions of unnecessary healthcare dollars will be spent in the United States as a result of insufficiently treating men with low testosterone. The FDA rules will only accelerate those huge expenses by creating a fear-mongering environment.

When the FDA went after drug manufacturers and their "Low T" campaign, the agency removed the drugmakers' ability to push public awareness of low testosterone. Certainly, there may be some valid criticism of the direct-to-consumer marketing tactics the drug companies used, but the campaign worked—testosterone prescriptions were remarkably on the rise during that period. When the new FDA rules came into place, however, that pace largely decelerated because physicians and patients misinterpreted the FDA's statements and rules.

FDA rules govern only manufacturers, but patients or doctors may think those rules regulate them, too. They don't. The FDA's requirement that drugmakers label testosterone as potentially causing increases in cardiovascular events and strokes leads people to believe that maybe those risks are real. They aren't. The FDA's stance is not at all based on science and has been disproven; to the contrary, the opposite has been repeatedly shown to be true—testosterone replacement is cardioprotective. Still, the FDA retains its rules, so people may believe those misstated risks are real.

Physicians have considerable fears of lawsuits related to cancers, and there is a mistaken belief that perhaps testosterone can increase the risk of a man getting prostate cancer. This has been completely disproven, but it's still a "legacy" thought, so to speak. I've asked physicians who have no role in testosterone replacement what their thoughts are regarding testosterone replacement and prostate cancer. These are very intelligent physicians, but again, they are not specialists in hormone replacement. They also practice traditional medicine in the sense that they adhere to what they were taught 20 years ago instead of keeping abreast of the latest studies and evidence. These physicians told me that they believe testosterone replacement would increase a man's chance of prostate cancer. When I sent them the science showing that TRT actually has no impact on prostate cancer, they were surprised they hadn't known that, because the information was so widespread and has been around for so long.

Unfortunately, some doctors still believe this inaccurate notion about there being a link between testosterone replacement and prostate cancer.

## What about side effects?

There are numerous misconceptions about testosterone replacement therapy. In this book, I've talked about the actual science behind TRT, not just my opinion of it. Obviously, I'm of the opinion that testosterone replacement therapy in testosterone-deficient men is an important health concern. My opinion is based on the science, not on my thoughts, beliefs, guesses, or gut instincts. I do not practice medicine primarily based on my gut. I certainly use instinct and gut feelings to guide decision making, yes—this is what we are supposed to do as physicians—but I do not allow my gut feelings or instincts to trump logic and evidence-based medicine.

Men have lots of questions about TRT, such as "What happens if I don't get enough testosterone from a pellet?" and "What if I get too much testosterone?" Surprisingly, the former situation happens very infrequently. In my practice, there have been a few occasions where the testosterone level did not rise as much as we expected. In such instances, we had the man take a shot of testosterone weekly to augment the pellets, and then we increased the pellet dose starting with the next round.

As far as the too-much scenario goes, I've never encountered a situation with a weight-based dosage resulting in too much testosterone, negative side effects, or abnormally high lab results. If that did occur, however, the answer would be to decrease the dose the next time. There would be no reason to remove a pellet since there are not many side effects of too much testosterone. About the only thing that can happen is a little bit of agitation, but that doesn't generally occur. It is true that the side effects of anabolic steroid abuse can be substantial, which is what men may be thinking of when they ask about what happens with too much testosterone. In general, though, side effects just don't occur at therapeutic testosterone replacement levels.

When a condition called polycythemia occurs, the body makes too many red blood cells. This has been noted to happen occasionally with testosterone replacement therapy. If it does occur, you need to donate blood to get rid of those extra red blood cells. It's typical to monitor a man's CBC (complete blood count) once or twice during the year when initiating testosterone.

A side effect associated with anabolic steroids in bodybuilders is the growth of breast tissue, which is obviously undesirable. This is pretty unheard of with therapeutic testosterone replacement therapy. Even though the likelihood of that happening is vanishingly small, it's a common practice to measure a man's estrogen level to see whether it has risen during TRT. (Testosterone and estrogen can each turn into each other.) Growth of breast tissue is associated with taking excess anabolic steroids or excess testosterone, which is not the purpose of testosterone replacement therapy.

One risk testosterone replacement *does* have is that infertility can develop. This is a real risk. If you're planning to have children in the future and you have low testosterone, then you need to discuss this with your doctor. Typically, men who plan to have children will take something called hCG to continue sperm and natural testosterone production. There's also evidence that other medications can be used to preserve fertility when taking testosterone. If testosterone is taken for only a year or so, fertility issues are generally not a concern. However, men taking testosterone because they have low testosterone due to no specific cause will likely take it indefinitely, so fertility issues need to be addressed.

## Hormone replacement at Allure

If you choose to visit my office, you'll be greeted by our a member of our First Impression Team, who will give you a tour of our office, as well as a welcome gift, and settle you into a patient room. The medical assistant will go over your concerns, answer any questions you might have, and obtain information about your basic health. The assistant also gives you a questionnaire to fill out and will introduce you to the physician assistant or doctor, who is specially trained in hormone replacement therapy for men. Typically, you get a prescription to have some blood work done.

It usually takes about two weeks to get the blood work back, so follow-up appointments are scheduled after that time. During the next visit, we go over your lab results and your symptoms, find any connected or underlying issues you may have, and—if appropriate—prescribe the necessary dosage of testosterone for you. (Typically, that's 10 mg of testosterone for every pound of body weight.) The testosterone pellets come in 200 mg increments and are very small—each one is the size of a grain of rice—so a man who weighs 200 pounds might get 2,000 mg of testosterone, which is 10 pellets. We then write the prescription for your follow-up blood work to check your testosterone and other hormone levels after a certain interval of time.

For updates on testosterone replacement therapy, to explore new things we are learning about testosterone replacement therapy, and to find out how we are potentially changing our practice with regard to testosterone replacement therapy, please visit our website at alluremedical.com. You can also schedule an appointment on our website, or you can call us at 586-992-8300.

It is my pleasure to serve you,

Dr. Charles Mok

# About the Author

**D**r. Charles Mok has been practicing medicine for over twenty-five years. After receiving his medical degree, Dr. Mok completed his post-graduate training in emergency medicine and became the vice chair of the emergency department at Mt. Clemens Hospital in Michigan, now known as McLaren Macomb. He saw countless patients with health emergencies that were fully preventable. In 2003, Dr. Mok founded Allure Medical, one of the largest and most successful practices of its kind in the country. Through Allure, he transitioned to helping patients with disease prevention. Allure is also a leader in elective cosmetic medical and surgical treatments as well as treating varicose veins, hair loss, and stem cell therapy. There are many misunderstood options for natural hormone replacement. Dr. Mok's mission is to present reliable research in a simple fashion so you can make the most effective decision for your health. He is also the author of *Testosterone, Strong Enough for a Man, Made for a Woman.*

# Bibliography

Akinloye, O., B. B. Popoola, M. B. Ajadi, J. G. Uchechukwu, and D. P. Oparinde. "Hypogonadism and Metabolic Syndrome in Nigerian Male Patients with Both Type 2 Diabetes and Hypertension." *International Journal of Endocrinology and Metabolism* 12, no. 1 (2014): e10749. doi:10.5812/ijem.10749.

Almehmadi, Y., A. A. Yassin, J. E. Nettlesip, and F. Saad. "Testosterone Replacement Therapy Improves the Health-Related Quality of Life of Men Diagnosed with Late-Onset Hypogonadism." *Arab Journal of Urology* 14 (2016): 31–36.

Andriole, G. L., E. D. Crawford, R. L. Grubb III, S. S. Buys, D. Chia, T. R. Church, M. N. Fouad, et al., for the PLCO Project Team. "Prostate Cancer Screening in the Randomized Prostate, Lung, Colorectal, and Ovarian Cancer Screening Trial: Mortality Results after 13 Years of Follow-Up." *Journal of the National Cancer Institute* 104, no. 2 (2012): 125–132. doi:10.1093/jnci/djr500.

Arver, S., B. Luong, A. Fraschke, O. Ghatnekar, S. Stanisic, D. Gultyev, and E. Müller. "Is Testosterone Replacement Therapy in Males with Hypogonadism Cost-Effective? An Analysis in Sweden." *Journal of Sexual Medicine* 11, no. 1 (2014): 262–272. doi:10.1111/jsm.12277.

Baas, W., and T. S. Köhler. "Testosterone Replacement Therapy and Voiding Dysfunction." *Translational Andrology and Urology* 5, no. 6 (2016): 890–897.

Baillargeon, J., Y.-F. Kuo, X. Fang, and V. B. Shahinian. "Long-Term Exposure to Testosterone Therapy and the Risk of High Grade Prostate Cancer." *Journal of Urology* 194 (2015): 1612–1616. doi:10.1016/j.juro.2015.05.099.

Baillargeon, J., R. J. Urban, Y.-F. Kuo, K. J. Ottenbacher, M. A. Raji, F. Du, Y.-L. Lin, and J. S. Goodwin. "Risk of Myocardial Infarction in Older Men Receiving Testosterone Therapy." *Annals of Pharmacotherapy* 48, no. 9 (2014): 1138–1144. doi:10.1177/1060028014539918.

Bain, J. "Testosterone and the Aging Male: To Treat or Not to Treat?" *Maturitas* 66 (2010): 16–22.

Basaria, S., S. M. Harman, T. G. Travison, H. Hodis, P. Tsitouras, M. Budoff, K. M. Pencina, et al. "Effects of Testosterone Administration for 3 Years on Subclinical Atherosclerosis Progression in Older Men with Lower Low-Normal Testosterone Levels: A Randomized Clinical Trial." *Journal of the American Medical Association* 314, no. 6 (2015): 570–581. doi:10.1001/jama.2015.8881.

Bhasin, S. "Effects of Testosterone Administration on Fat Distribution, Insulin Sensitivity, and Atherosclerosis Progression." *Clinical Infectious Diseases* 37, suppl. 2 (2003): S142–S149.

Bhattacharya, R. K., M. Khera, G. Blick, H. Kushner, and M. M. Miner. "Testosterone Replacement Therapy among Elderly Males: The Testim Registry in the US (TRiUS)." *Aging* 7 (2012): 321–330.

Białek, M., P. Zaremba, K. K. Borowicz, and S. J. Czuczwar. "Neuroprotective Role of Testosterone in the Nervous System." *Polish Journal of Pharmacology* 56 (2004): 509–518.

Blue Cross Blue Shield Blue Care Network of Michigan. "Medication Use Policy: Testosterone Replacement Therapy." Specialty Workgroup Meeting, Sept. 21, 2015.

Blue Cross Blue Shield Blue Care Network of Michigan. "Prior Authorization Medical Coverage Drug List."

Borst, S. E., C. F. Conover, C. S. Carter, C. M. Gregory, E. Marzetti, C. Leeuwenburgh, K. Vandenborne, and T. J. Wronski. "Anabolic Effects of Testosterone Are Preserved during Inhibition of 5-Reductase." *American Journal of Physiology Endocrinology and Metabolism* 293 (2007): E507–E514. doi:10.1152/ajpendo.00130.2007.

Borst, S. E., J. J. Shuster, B. Zou, F. Ye, H. Jia, A. Wokhlu, and J. F. Yarrow. "Cardiovascular Risks and Elevation of Serum DHT Vary by Route of Testosterone Administration: A Systematic Review and Meta-Analysis." *BMC Medicine* 12 (2014): 211.

Borst, S. E., and J. F. Yarrow. "Injection of Testosterone May Be Safer and More Effective than Transdermal Administration for Combating Loss of Muscle and Bone in Older Men." *American Journal of Physiology: Endocrinology and Metabolism* 308 (2015): E1035–E1042. doi:10.1152/ajpendo.00111.2015.

Boyanov, M. A., Z. Boneva, and V. G. Christov. "Testosterone Supplementation in Men with Type 2 Diabetes, Visceral Obesity and Partial Androgen Deficiency." *Aging Male* 6, no. 1 (2003): 1–7. doi:10.1080/tam.6.1.1.7.

Brawer, M. K. "Androgen Supplementation and Prostate Cancer Risk: Strategies for Pretherapy Assessment and Monitoring." *Reviews in Urology* 5, suppl. 1 (2003): S29–S33.

Brock, G., D. Heiselman, M. Maggi, S. W. Kim, J. M. Rodríguez Vallejo, H. M. Behre, J. McGettigan, et al. "Effect of Testosterone Solution 2% on Testosterone Concentration, Sex Drive and Energy in Hypogonadal Men: Results of a Placebo Controlled Study." *Journal of Urology* 195 (2016): 699–705. doi:10.1016/j.juro.2015.10.083.

Burris, A. S., S. M. Banks, C. S. Carter, J. M. Davidson, and R. J. Sherins. "A Long-Term, Prospective Study of the Physiologic and Behavioral Effects of Hormone Replacement in Untreated Hypogonadal Men." *Journal of Andrology* 13, no. 4 (1992): 297–304.

Cai, X., Y. Tian, T. Wu, C.-X. Cao, H. Li, and K.-J. Wang. "Metabolic Effects of Testosterone Replacement Therapy on Hypogonadal Men with Type 2 Diabetes Mellitus: A Systematic Review and Meta-Analysis of Randomized Controlled Trials." *Asian Journal of Andrology* 16 (2014): 146–152.

Carnegie, C. "Diagnosis of Hypogonadism: Clinical Assessments and Laboratory Tests." *Reviews in Urology* 6, suppl. 6 (2004): S3–S8.

Carruthers, M., P. Cathcart, and M. R. Feneley. "Evolution of Testosterone Treatment over 25 Years: Symptom Responses, Endocrine Profiles and Cardiovascular Changes." *Aging Male* 18, no. 4 (2015): 217–227. doi:10.3109/13685538.2015.1048218.

Cherrier, M. M., S. Craft, and A. H. Matusmoto. "Cognitive Changes Associated with Supplementation of Testosterone or Dihydrotestosterone in Mildly Hypogonadal Men: A Preliminary Report." *Journal of Andrology* 24, no. 4 (2003): 566–576.

Cherrier, M. M., A. M. Matsumoto, J. K. Amory, S. Asthana, W. Bremner, E. R. Peskind, M. A. Raskind, and S. Craft. "Testosterone Improves Spatial Memory in Men with Alzheimer Disease and Mild Cognitive Impairment." *Neurology* 64 (2005): 2063–2068.

Chueh, K.-S., S.-P. Huang, Y.-C. Lee, C.-J. Wang, H.-C. Yeh, W.-M. Li, W.-J. Wu, et al. "The Comparison of the Aging Male Symptoms (AMS) Scale and Androgen Deficiency in the Aging Male (ADAM) Questionnaire to Detect Androgen Deficiency in Middle-Aged Men." *Journal of Andrology* 33, no. 5 (2012): 817–823.

Conaglen, H. M., R. G. Paul, T. Yarndley, J. Kamp, M. S. Elston, and J. V. Conaglen. "Retrospective Investigation of Testosterone Undecanoate Depot for the Long-Term Treatment of Male Hypogonadism in Clinical Practice." *Journal of Sexual Medicine* 11, no. 2 (2014): 574–582.

Corona, G., S. Bianchini, A. Sforza, L. Vignozzi, and M. Maggi. "Hypogonadism as a Possible Link between Metabolic Diseases and Erectile Dysfunction in Aging Men." *Hormones* 14, no. 4 (2015): 569–578.

Corona, G., V. A. Giagulli, E. Maseroli, L. Vignozzi, A. Aversa, M. Zitzmann, F. Saad, E. Mannucci, and M. Maggi. "Testosterone Supplementation and Body Composition: Results from a Meta-Analysis Study." *European Journal of Endocrinology* 174, no. 3 (2016): R99–R116.

Corona, G., M. Monami, G. Rastrelli, A. Aversa, A. Sforza, A. Lenzi, G. Forti, E. Mannucci, and M. Maggi. "Type 2 Diabetes Mellitus and Testosterone: A Meta-Analysis Study." *International Journal of Andrology* 34 (2010): 528–540.

Corona, G., G. Rastrelli, E. Maseroli, N. Fralassi, A. Sforza, G. Forti, E. Mannucci, and M. Maggi. "Low Testosterone Syndrome Protects Subjects with High Cardiovascular Risk Burden from Major Adverse Cardiovascular Events." *Andrology* 2 (2014): 741–747. doi:10.1111/j.2047-2927.2014.00241.x.

Corona, G., G. Rastrelli, E. Maseroli, A. Sforza, and M. Maggi. "Testosterone Replacement Therapy and Cardiovascular Risk: A Review." *World Journal of Men's Health* 33, no. 3 (2015): 130–142. doi:10.5534/wjmh.2015.33.3.130.

Corona, G., G. Rastrelli, M. Monami, F. Saad, M. Luconi, M. Lucchese, E. Facchiano, et al. "Body Weight Loss Reverts Obesity-Associated Hypogonadotropic Hypogonadism: A Systematic Review and Meta-Analysis." *European Journal of Endocrinology* 168 (2013): 829–843.

Corona, G., L. Vignozzi, A. Sforza, and M. Maggi. "Risks and Benefits of Late Onset Hypogonadism Treatment: An Expert Opinion." *World Journal of Men's Health* 31, no. 2 (2013): 103–125. doi:10.5534/wjmh.2013.31.2.103.

Coward, R. M., J. Simhan, and C. C. Carson III. "Prostate-Specific Antigen Changes and Prostate Cancer in Hypogonadal Men Treated with Testosterone Replacement Therapy." *BJU International* 103 (2008): 1179–1183. doi:10.1111/j.1464-410X.2008.08240.x.

Crawford, D., W. Poage, A. Nyhuis, D. A. Price, S. A. Dowsett, S. Gelwicks, and D. Muram. "Measurement of Testosterone: How Important Is a Morning Blood Draw?" *Current Medical Research and Opinion* 31, no. 10 (2015): 1911–1914. doi:10.1185/03007995.2015.1082994.

Crawford, M., and L. Kennedy. "Testosterone Replacement Therapy: Role of Pituitary and Thyroid in Diagnosis and Treatment." *Translational Andrology and Urology* 5, no. 6 (2016): 850–858.

Crosnoe, L. E., E. Grober, D. Ohl, and E. D. Kim. "Exogenous Testosterone: A Preventable Cause of Male Infertility." *Translational Andrology and Urology* 2 (2013): 106–113. doi:10.3978/j.issn.2223-4683.2013.06.01.

Cui, Y., H. Zong, H. Yan, and Y. Zhang. "The Effect of Testosterone Replacement Therapy on Prostate Cancer: A Systematic Review and Meta-Analysis." *Prostate Cancer and Prostatic Diseases* 17, no. 2 (2014): 132–143. doi:10.1038 /pcan.2013.60.

Cunningham, G. R. "Testosterone and Metabolic Syndrome." *Asian Journal of Andrology* 17 (2015): 192–196.

Daig, I., L. A. J. Heinemann, S. Kim, S. Leungwattanakij, X. Badia, E. Myon, C. Moore, F. Saad, P. Potthoff, and D. M. Thai. "The Aging Males' Symptoms (AMS) Scale: Review of Its Methodological Characteristics." *Health and Quality of Life Outcomes* 1 (2003): 77.

Desroches, B., T. P. Kohn, C. Welliver, and A. W. Pastuszak. "Testosterone Therapy in the New Era of Food and Drug Administration Oversight." *Translational Andrology and Urology* 5, no. 2 (2016): 207–212.

Dias, J. P., D. Melvin, E. M. Simonsick, O. Carlson, M. D. Shardell, L. Ferrucci, C. W. Chia, S. Basaria, and J. M. Egan. "Effects of Aromatase Inhibition vs. Testosterone in Older Men with Low Testosterone: Randomized Controlled Trial." *Andrology* 4 (2016): 33–40. doi:10.1111/andr.12126.

DiGiorgio, L., and H. Sadeghi-Nejad. "Off Label Therapies for Testosterone Replacement." *Translational Andrology and Urology* 5, no. 6 (2016): 844–849. doi:10.21037/tau.2016.08.15.

Di Loreto, C., F. La Marra, G. Mazzon, E. Belgrano, C. Trombetta, and S. Cauci. "Immunohistochemical Evaluation of Androgen Receptor and Nerve Structure Density in Human Prepuce from Patients with Persistent Sexual Side Effects after Finasteride Use for Androgenetic Alopecia." *PLoS ONE* 9, no. 6 (2014): e100237. doi:10.1371/journal.pone.0100237.

Dimopouloua, C., I. Ceausub, H. Depyperec, I. Lambrinoudakid, A. Muecke, F. R. Pérez-Lópezf, M. Reesg, et al. "EMAS Position Statement: Testosterone Replacement Therapy in the Aging Male." *Maturitas* 84 (2016): 94–99.

Drew, T., D. Olszewska-Seonina, and P. Chlosta. "Testosterone Replacement Therapy in Obese Males." *Acta Poloniae Pharmaceutica Drug Research* 68, no. 5 (2011): 623–627.

Ebrahimi, F., and M. Christ-Crain. "Metabolic Syndrome and Hypogonadism: Two Peas in a Pod." *Swiss Medal Weekly* 146 (2016): w14283.

Eisenberg, M. L. "Testosterone Replacement Therapy and Prostate Cancer Incidence." *World Journal of Men's Health* 33, no. 3 (2015): 125–129. doi:10.5534/wjmh.2015.33.3.125.

Eisenberg, M. L., S. Li, P. Betts, D. Herder, D. J. Lamb, and L. I. Lipshultz. "Testosterone Therapy and Cancer Risk." *BJU International* 115 (2015): 317–321. doi:10.1111/bju.12756.

Eisenberg, M. L., S. Li, D. Herder, D. J. Lamb, and L. I. Lipshultz. "Testosterone Therapy and Mortality Risk." *International Journal of Impotence Research* 27, no. 2 (2015): 46–48. doi:10.1038/ijir.2014.29.

Etminan, M., S. C. Skeldon, S. L. Goldenberg, B. Carleton, and J. M. Brophy. "Testosterone Therapy and Risk of Myocardial Infarction: A Pharmacoepidemiologic Study." *Pharmacotherapy* 35, no. 1 (2015): 72–78. doi:10.1002/phar.1534.

Endo Pharmaceuticals. "Testopel (Testosterone Pellets)." Revised October 2016.

Finkle, W. D., S. Greenland, G. K. Ridgeway, J. L. Adams, M. A. Frasco, M. B. Cook, J. F. Fraumeni Jr., and R. N. Hoover. "Increased Risk of Non-Fatal Myocardial Infarction Following Testosterone Therapy Prescription in Men." *PLoS ONE* 9, no. 1 (2014): e85805.

Francomano, D., R. Bruzziches, G. Barbaro, A. Lenzi, and A. Aversa. "Effects of Testosterone Undecanoate Replacement and Withdrawal on Cardio-metabolic, Hormonal and Body Composition Outcomes in Severely Obese Hypogonadal Men: A Pilot Study." *Journal of Endocrinological Investigation* 37 (2014): 401–411. doi:10.1007/s40618-014-0066-9.

Francomano, D., A. Ilacqua, R. Bruzziches, A. Lenzi, and A. Aversa. "Effects of 5-Year Treatment with Testosterone Undecanoate on Lower Urinary Tract Symptoms in Obese Men with Hypogonadism and Metabolic Syndrome." *Urology* 83 (2014): 167e174.

Frederiksen, L., K. Højlund, D. M. Hougaard, T. H. Mosbech, R. Larsen, A. Flyvbjerg, J. Frystyk, K. Brixen, and M. Andersen. "Testosterone Therapy Decreases Subcutaneous Fat and Adiponectin in Aging Men." *European Journal of Endocrinology* 16 (2012): 469–476.

Fui, M. N. T., P. Dupuis, and M. Grossmann. "Lowered Testosterone in Male Obesity: Mechanisms, Morbidity and Management." *Asian Journal of Andrology* 16 (2014): 223–231.

Fui, M. N. T., L. A. Prendergast, P. Dupuis, M. Raval, B. J. Strauss, J. D. Zajac, and M. Grossmann. "Effects of Testosterone Treatment on Body Fat and Lean Mass in Obese Men on a Hypocaloric Diet: A Randomised Controlled Trial." *BMC Medicine* 14 (2016): 153. doi:10.1186/s12916-016-0700-9.

Fujioka, K. "Current and Emerging Medications for Overweight or Obesity in People with Comorbidities." *Diabetes, Obesity and Metabolism* (2015) 17: 1021–1032.

Golan, R., J. M. Scovell, and R. Ramasamy. "Age-Related Testosterone Decline Is Due to Waning of Both Testicular and Hypothalamic-Pituitary Function." *Aging Male* 13, no. 3 (2015): 201–204. doi:10.3109/13685538.2015.10523 92.

Goodman, N., A. Guay, P. Dandona, S. Dhindsa, C. Faiman, G. R. Cunningham, for the AACE Reproductive Endocrinology Scientific Committee. "American Association of Clinical Endocrinologists and American College of Endocrinology Position Statement on the Association of Testosterone and Cardiovascular Risk." *Endocrine Practice* 21, no. 9 (2015): 1066–1073.

Gooren, L. J. G., and F. Saad. "Recent Insights into Androgen Action on the Anatomical and Physiological Substrate of Penile Erection." *Asian Journal of Andrology* 8, no. 1 (2006): 3–9. doi:10.1111/j.1745-7262.2006.00105.x.

Gould, D. C., M. R. Feneley, and R. S. Kirby. "Prostate-Specific Antigen Testing in Hypogonadism: Implications for the Safety of Testosterone Replacement Therapy." *BJU International* 98 (2006): 1–9 doi:10.1111/j.1464-410X.2006. 06191,06242,06243,06245.x.

Gouras, G. K., H. Xu, R. S. Gross, J. P. Greenfield, B. Hai, R. Wang, and P. Greengard. "Testosterone Reduces Neuronal Secretion of Alzheimer's B-amyloid Peptides." *Proceedings of the National Academy of Sciences* 97, no. 3 (2000): 1202–1205.

Grabner, M., A. Bodhani, N. Khandelwal, S. Palli, N. Bonine, and M. Khera. "Clinical Characteristics, Health Care Utilization and Costs among Men with Primary or Secondary Hypogonadism in a US Commercially Insured Population." *Journal of Sexual Medicine* 14 (2017): 88–97.

Grech, A., J. Breck, and J. Heidelbaugh. "Adverse Effects of Testosterone Replacement Therapy: An Update on the Evidence and Controversy." *Therapeutic Advances in Drug Safety* 5, no. 5 (2014): 190–200. doi:10.1177 /2042098614548680.

Guo, C., W. Gu, M. Liu, B. Peng, X. Yao, B. Yang, and J. Zheng. "Efficacy and Safety of Testosterone Replacement Therapy in Men with Hypogonadism: A Meta-Analysis Study of Placebo-Controlled Trials." *Experimental and Therapeutic Medicine* 11 (2016): 853–863.

Hackett, G. "An Update on the Role of Testosterone Replacement Therapy in the Management of Hypogonadism." *Therapeutic Advances in Urology* 8, no. 2 (2016): 147–160. doi:10.1177/1756287215617648.

Hackett, G., N. Cole, A. Saghir, P. Jones, R. C. Strange, and S. Ramachandran. "Testosterone Undecanoate Improves Sexual Function in Men with Type 2 Diabetes and Severe Hypogonadism: Results from a 30-Week Randomized Placebo-Controlled Study." *BJU International* 118, no. 5 (2016): 804–813. doi:10.1111/bju.13516.

Hackett, G., P. W. Jones, R. C. Strange, and S. Ramachandran. "Statin, Testosterone and Phosphodiesterase 5-Inhibitor Treatments and Age Related Mortality in Diabetes." *World Journal of Diabetes* 8, no. 3 (2017): 104–111. doi:10.4239/wjd.v8.i3.104.

Hackett, G., M. Kirby, and A. J. Sinclair. "Testosterone Deficiency, Cardiac Health, and Older Men." *International Journal of Endocrinology* 2014 (2014): 143763. doi:10.1155/2014/143763.

Haddad, R. M., C. C. Kennedy, S. M. Caples, M. J. Tracz, E. R. Bolona, K. Sideras, M. V. Uraga, P. J. Erwin, and V. M. Montori. "Testosterone and Cardiovascular Risk in Men: A Systematic Review and Meta-Analysis of Randomized Placebo-Controlled Trials." *Mayo Clinic Proceedings* 82, no. 1 (2007): 29–39.

Haider, A., L. J. Gooren, P. Padungtod, and F. Saad. "Beneficial Effects of 2 Years of Administration of Parenteral Testosterone Undecanoate on the Metabolic Syndrome and on Non-alcoholic Liver Steatosis and C-Reactive Protein." *Hormone Molecular Biology and Clinical Investigation* 1, no. 1 (2010): 27–33. doi:10.1515/HMBCI.2010.002.

Haider, A., F. Saad, G. Doros, and L. Gooren. "Hypogonadal Obese Men with and without Diabetes Mellitus Type 2 Lose Weight and Show Improvement in Cardiovascular Risk Factors When Treated with Testosterone: An Observational Study." *Obesity Research and Clinical Practice* 8 (2014): e339–e349.

Haider, A., A. Yassin, G. Doros, and F. Saad. "Effects of Long-Term Testosterone Therapy on Patients with 'Diabesity': Results of Observational Studies of Pooled Analyses in Obese Hypogonadal Men with Type 2 Diabetes." *International Journal of Endocrinology* 2014 (2014): 683515. doi:10.1155/2014/683515.

Haider, A., A. Yassin, K. S. Haider, G. Doros, F. Saad, and G. M. C. Rosano. "Men with Testosterone Deficiency and a History of Cardiovascular Diseases Benefit from Long-Term Testosterone Therapy: Observational, Real-Life Data from a Registry Study." *Vascular Health and Risk Management* 12 (2016): 251–261.

Hall, J. R., A. R. Wiechmann, R. L. Cunningham, L. A. Johnson, M. Edwards, R. C. Barber, M. Singh, S. Winter, and S. E. O'Bryant, for the Texas Alzheimer's Research and Care Consortium. "Total Testosterone and Neuropsychiatric Symptoms in Elderly Men with Alzheimer's Disease." *Alzheimer's Research and Therapy* 7 (2015): 24. doi:10.1186/s13195-015-0107-4.

Harada, N., R. Hanaoka, K. Hanada, T. Izawa, H. Inui, and R. Yamaji. "Hypogonadism Alters Cecal and Fecal Microbiota in Male Mice." *Gut Microbes* 7, no. 6 (2016): 533–539. doi:10.1080/19490976.2016.1239680.

Hassan, J., and J. Barkin. "Testosterone Deficiency Syndrome: Benefits, Risks, and Realities Associated with Testosterone Replacement Therapy." *Canadian Journal of Urology* 23, suppl. 1 (2016): 20–30.

He, J., S. Bhasin, E. F. Binder, K. E. Yarasheski, C. Castaneda-Sceppa, E. T. Schroeder, R. Roubenoff, C.-P. Chou, S. P. Azen, and F. R. Sattler. "Cardiometabolic Risks during Anabolic Hormone Supplementation in Older Men." *Obesity* 21, no. 5 (2013): 968–975. doi:10.1002/oby.20081.

Heinemann, L. A., C. Moore, J. C. Dinger, and D. Stoehr. "Sensitivity as Outcome Measure of Androgen Replacement: The AMS Scale." *Health and Quality of Life Outcomes* 4 (2006): 23. doi:10.1186/1477-7525-4-23.

Ho, C. C. K., S. F. Tong, W. Y. Low, C. J. Ng, E. M. Khoo, V. K. M. Lee, Z. M. Zainuddin, and H. M. Tan. "A Randomized, Double-Blind, Placebo-Controlled Trial on the Effect of Long-Acting Testosterone Treatment as Assessed by the Aging Male Symptoms Scale." *BJU International* 110 (2011): 260–265. doi:10.1111/j.1464-410X.2011.10755.x.

Høst, C., L. C. Gormsen, B. Christensen, N. Jessen, D. M. Hougaard, J. S. Christiansen, S. B. Pedersen, M. D. Jensen, S. Nielsen, and C. H. Gravholt. "Independent Effects of Testosterone on Lipid Oxidation and VLDL-TG Production: A Randomized, Double-Blind, Placebo-Controlled, Crossover Study." *Diabetes* 62 (2013): 1409–1416.

Hua, J. T., K. L. Hildreth, and V. S. Pelak. "Effects of Testosterone Therapy on Cognitive Function in Aging: A Systematic Review." *Cognitive Behavior and Neurology* 29 (2016): 122–138.

Hussain, R., A. M. Ghoumari, B. Bielecki, J. Steibel, N. Boehm, P. Liere, W. B. Macklin, et al. "The Neural Androgen Receptor: A Therapeutic Target for Myelin Repair in Chronic Demyelination." *Brain* 136 (2013): 132–146.

Hwang, K., and M. Miner. "Controversies in Testosterone Replacement Therapy: Testosterone and Cardiovascular Disease." *Asian Journal of Andrology* 17 (2015): 187–191.

Ip, F. F., I. di Pierro, R. Brown, I. Cunningham, D. J. Handelsman, and P. Y. Liu. "Trough Serum Testosterone Predicts the Development of Polycythemia in Hypogonadal Men Treated for up to 21 Years with Subcutaneous Testosterone Pellets." *European Journal of Endocrinology* 162 (2010): 385–390.

Isidori, A. M., G. Balercia, A. E. Calogero, G. Corona, A. Ferlin, S. Francavilla, D. Santi, and M. Maggi. "Outcomes of Androgen Replacement Therapy in Adult Male Hypogonadism: Recommendations from the Italian Society of Endocrinology." *Journal of Endocrinological Investigation* 38 (2015): 103–112. doi:10.1007/s40618-014-0155-9.

Janjgava, S., T. Zerekidze, L. Uchava, E. Giorgadze, and K. Asatiani. "Influence of Testosterone Replacement Therapy on Metabolic Disorders in Male Patients with Type 2 Diabetes Mellitus and Androgen Deficiency." *European Journal of Medical Research* 19 (2014): 56.

Jeong, S. M., B. K. Ham, M. G. Park, M. M. Oh, D. K. Yoon, J. J. Kim, and D. G. Moon. "Effect of Testosterone Replacement Treatment in Testosterone Deficiency Syndrome Patients with Metabolic Syndrome." *Korean Journal of Urology* 52 (2011): 566–571. doi:10.4111/kju.2011.52.8.566.

Jia, H., C. T. Sullivan, S. C. McCoy, J. F. Yarrow, M. Morrow, and S. E. Borst. "Review of Health Risks of Low Testosterone and Testosterone Administration." *World Journal of Clinical Cases* 3, no. 4 (2015): 338–344.

Jones, T. H., S. Arver, H. M. Behre, J. Buvat, E. Meuleman, I. Moncada, A. M. Morales, et al., and TIMES2 Investigators. "Testosterone Replacement in Hypogonadal Men with Type 2 Diabetes and/or Metabolic Syndrome (the TIMES2 Study)." *Diabetes Care* 34 (2011): 828–837.

Jung, J. H., S. V. Ahn, J. M. Song, S.-J. Chang, K. J. Kim, S. W. Kwon, S.-Y. Park, and S.-B. Koh. "Obesity as a Risk Factor for Prostatic Enlargement: A Retrospective Cohort Study in Korea." *International Neurology Journal* 20 (2016): 321–328. doi:10.5213/inj.1632584.292.

Jung, J. H., and H. S. Shin. "Effect of Testosterone Replacement Therapy on Cognitive Performance and Depression in Men with Testosterone Deficiency Syndrome." *World Journal of Men's Health* 34, no. 3 (2016): 194–199. doi:10.5534/wjmh.2016.34.3.194.

Kanayama, G., J. I. Hudson, J. DeLuca, S. Isaacs, A. Baggish, R. Weiner, S. Bhasin, and H. G. Pope Jr. "Prolonged Hypogonadism in Males Following Withdrawal from Anabolic-Androgenic Steroids: An Underrecognized Problem." *Addiction* 110, no. 5 (2015): 823–831. doi:10.1111/add.12850.

Kapoor, D., E. Goodwin, K. S. Channer, and T. H. Jones. "Testosterone Replacement Therapy Improves Insulin Resistance, Glycaemic Control, Visceral Adiposity and Hypercholesterolaemia in Hypogonadal Men with Type 2 Diabetes." *European Journal of Endocrinology* 154 (2006): 899–906.

Katayev, A., C. Balciza, and D. W. Seccombe. "Establishing Reference Intervals for Clinical Laboratory Test Results: Is There a Better Way?" *American Journal of Clinical Pathology* 133 (2010): 180–186. doi:10.1309/AJCPN5BMTSF1CDYP.

Katz, A., A. Katz, and C. Burchill. "Androgen Therapy: Testing before Prescribing and Monitoring during Therapy." *Canadian Family Physician* 53 (2007): 1936–1942.

Kava, B. R. "To Treat or Not to Treat with Testosterone Replacement Therapy: A Contemporary Review of Management of Late-Onset Hypogonadism and Critical Issues Related to Prostate Cancer." *Current Urology Reports* 15 (2014): 422. doi:10.1007/s11934-014-0422-5.

Kelleher, S., A. J. Conway, and D. J. Handelsman. "A Randomised Controlled Clinical Trial of Antibiotic Impregnation of Testosterone Pellet Implants to Reduce Extrusion Rate." *European Journal of Endocrinology* 146 (2002): 513–518.

Kelleher, S., C. Howe, A. J. Conway, and D. J. Handelsman. "Testosterone Release Rate and Duration of Action of Testosterone Pellet Implants." *Clinical Endocrinology* 60 (2004): 420–428. doi:10.1111/j.1365-2265.2004.01994.x.

Khera, M. "Patients with Testosterone Deficit Syndrome and Depression." *Archives of Spanish Urology* 66, no. 7 (2013): 72936.

Kim, C., E. Barrett-Connor, V. R. Aroda, K. J. Mather, C. A. Christophi, E. S. Horton, X. Pi-Sunyer, G. A. Bray, F. Labrie, and S. H. Golden, on behalf of the Diabetes Prevention Program Research Group. "Testosterone and Depressive Symptoms among Men in the Diabetes Prevention Program." *Psychoneuroendocrinology* 72 (2016): 63–71. doi:10.1016/j.psyneuen.2016.06.009.

Kische, H., S. Gross, H. Wallaschofski, H. Jorgen Grabe, H. Volzke, M. Nauck, and R. Haring. "Associations of Androgens with Depressive Symptoms and Cognitive Status in the General Population." *PLoS ONE* 12, no. 5 (2017): e0177272. doi:10.1371/journal.pone.0177272.

Ko, Y. H., D. G. Moon, and K. H. Moon. "Testosterone Replacement Alone for Testosterone Deficiency Syndrome Improves Moderate Lower Urinary Tract Symptoms: One Year Follow-Up." *World Journal of Men's Health* 31, no. 1 (2013): 47–52. doi:10.5534/wjmh.2013.31.1.47.

Kohn, T. P., M. R. Louis, S. M. Pickett, M. C. Lindgren, J. R. Kohn, A. W. Pastuszak, and L. I. Lipshultz. "Age and Duration of Testosterone Therapy Predict Time to Return of Sperm Count after Human Chorionic Gonadotropin Therapy." *Fertility and Sterility* 107, no. 2 (2016): 351–357. doi:10.1016 /j.fertnstert.2016.10.004.

Køster-Rasmussen, R., M. K. Simonsen, V. Siersma, J. E. Henriksen, B. L. Heitmann, and N. de Fine Olivarius. "Intentional Weight Loss and Longevity in Overweight Patients with Type 2 Diabetes: A Population-Based Cohort Study." *PLoS ONE* 11, no. 1 (2016): e0146889. doi:10.1371 /journal.pone.0146889.

Kovac, J. R., S. Rajanahally, R. P. Smith, R. M. Coward, D. J. Lamb, and L. I. Lipshultz. "Patient Satisfaction with Testosterone Replacement Therapies: The Reasons behind the Choices." *Journal of Sexual Medicine* 11, no. 2 (2014): 553–562. doi:10.1111/jsm.12369.

Kurth, F., E. Luders, N. L. Sicotte, C. Gaser, B. S. Giesser, R. S. Swerdloff, M. J. Montag, R. R. Voskuh, and A. Mackenzie-Graham. "Neuroprotective Effects of Testosterone Treatment in Men with Multiple Sclerosis." *NeuroImage: Clinical 9* (2014): 454–460.

Layton, J. B., C. R. Meier, J. L. Sharpless, T. Stürmer, S. S. Jick, and M. A. Brookhart. "Comparative Safety of Testosterone Dosage Forms." *JAMA Internal Medicine* 175, no. 7 (2015): 1187–1196. doi:10.1001/jamainternmed.2015.1573.

Legros, J.-J., E. J. H. Meuleman, J. M. H. Elbers, T. B. P. Geurts, M. J. G. H. Kaspers, P. M. G. Bouloux, for the Study 43203 Investigators. "Oral Testosterone Replacement in Symptomatic Late-Onset Hypogonadism: Effects on Rating Scales and General Safety in a Randomized, Placebo-Controlled Study." *European Journal of Endocrinology* 160 (2009): 821–831.

Leibowitz, R. L., T. B. Dorff, S. Tucker, J. Symanowski, and N. J. Vogelzang. "Testosterone Replacement in Prostate Cancer Survivors with Hypogonadal Symptoms." *BJU International* 105 (2009): 1397–1401. doi:10.1111/j.1464-410X.2009.08980.x.

Majzoub, A., and E. Sabanegh Jr. "Testosterone Replacement in the Infertile Man." *Translational Andrology and Urology* 5, no. 6 (2016): 859–865.

Majzoub, A., and D. A. Shoskes. "A Case Series of the Safety and Efficacy of Testosterone Replacement Therapy in Renal Failure and Kidney Transplant Patients." *Translational Andrology and Urology* 5, no. 6 (2016): 814–818. doi:10.21037/tau.2016.07.09.

Malan, N. T., W. Smith, R. von Känel, M. Hamer, A. E. Schutte, and L. Malan. "Low Serum Testosterone and Increased Diastolic Ocular Perfusion Pressure: A Risk for Retinal Microvasculature." *Vasa* 44 (2015): 435–443. doi:10.1024/0301-1526/a000466.

Malkin, C. J., T. H. Jones, and K. S. Channer. "The Effect of Testosterone on Insulin Sensitivity in Men with Heart Failure." *European Journal of Heart Failure* 9 (2007): 44–50.

Malkin, C. J., P. J. Pugh, P. D. Morris, K. E. Kerry, R. D. Jones, T. H. Jones, and K. S. Channer. "Testosterone Replacement in Hypogonadal Men with Angina Improves Ischaemic Threshold and Quality of Life." *Heart* 90 (2004): 871–876. doi:10.1136/hrt.2003.021121.

Malkin, C. J., P. J. Pugh, J. N. West, E. J. R. van Beek, T. H. Jones, and K. S. Channer. "Testosterone Therapy in Men with Moderate Severity Heart Failure: A Double-Blind Randomized Placebo Controlled Trial." *European Heart Journal* 27 (2006): 57–64. doi:10.1093/eurheartj/ehi443.

Maneschi, F., A. Morelli, S. Filippi, I. Cellai, P. Comeglio, B. Mazzanti, T. Mello, et al. "Testosterone Treatment Improves Metabolic Syndrome-Induced Adipose Tissue Derangements." *Journal of Endocrinology* 215 (2012): 347–362.

Mathur, A., C. Malkin, B. Saeed, R. Muthusamy, T. H. Jones, and K. Channer. "Long-Term Benefits of Testosterone Replacement Therapy on Angina Threshold and Atheroma in Men." *European Journal of Endocrinology* 161 (2009): 443–449.

McCullough, A. "A Review of Testosterone Pellets in the Treatment of Hypogonadism." *Current Sexual Health Reports* 6 (2014): 265–269. doi:10.1007/s11930-014-0033-7.

McLaren, D., D. R. Siemens, J. Izard, A. Black, and A. Morales. "Clinical Practice Experience with Testosterone Treatment in Men with Testosterone Deficiency Syndrome." *BJU International* 102 (2008): 1142–1146. doi:10.1111/j.1464-410X.2008.07811.x.

Mechlin, C. W., J. Frankel, and A. McCullough. "Coadministration of Anastrozole Sustains Therapeutic Testosterone Levels in Hypogonadal Men Undergoing Testosterone Pellet Insertion." *Journal of Sexual Medicine* 11, no. 1 (2014): 254–261. doi:10.1111/jsm.12320.

Metzger, S. O., and A. L. Burnett. "Impact of Recent FDA Ruling on Testosterone Replacement Therapy (TRT)." *Translational Andrology and Urology* 5, no. 6 (2016): 921–926. doi:10.21037/tau.2016.09.08.

Michaud, J. E., K. L. Billups, and A. W. Partin. "Testosterone and Prostate Cancer: An Evidence-Based Review of Pathogenesis and Oncologic Risk." *Therapeutic Advances in Urology* 7, no. 6 (2015): 378–387. doi:10.1177/1756287215597633.

Millar, A. C., D. S. Elterman, L. Goldenberg, B. Van Asseldonk, A. Curtis, and K. Jarvi. "A Survey of Canadian Urologists' Opinions and Prescribing Patterns of Testosterone Replacement Therapy in Men on Active Surveillance for Low-Risk Prostate Cancer." *Canadian Urological Association Journal* 10, nos. 5–6 (2016): 181–184. doi:10.5489/cuaj.3608.

Morgentaler, A. "Testosterone Deficiency and Cardiovascular Mortality." *Asian Journal of Andrology* 17 (2015): 26–31.

Morgentaler, A., M. Zitzmann, A. M. Traish, A. W. Fox, T. H. Jones, M. Maggi, S. Arver, et al. "Fundamental Concepts Regarding Testosterone Deficiency and Treatment: International Expert Consensus Resolutions." *Mayo Clinic Proceedings* 91, no. 7 (2016): 881–896. doi:10.1016/j.mayocp.2016.04.007.

Morgentaler, A., et al. "Deaths and Cardiovascular Events in Men Receiving Testosterone: Comment and Response." *Journal of the American Medical Association* 311, no. 9 (2014): 961–965.

Morris, M. J., D. Huang, W. K. Kelly, S. F. Slovina, R. D. Stephenson, C. Eicher, A. Delacruz, T. Curley, L. Schwartz, and H. I. Scher. "Phase 1 Trial of High-Dose Exogenous Testosterone in Patients with Castration-Resistant Metastatic Prostate Cancer." *European Urology* 56, no. 2 (2009): 237–244. doi:10.1016/j.eururo.2009.03.073.

Morris, P. D., and K. S. Channer. "Testosterone and Cardiovascular Disease in Men." *Asian Journal of Andrology* 14 (2012): 428–435.

Muraleedharan, V., H. Marsh, D. Kapoor, K. S. Channer, and T. H. Jones. "Testosterone Deficiency Is Associated with Increased Risk of Mortality and Testosterone Replacement Improves Survival in Men with Type 2 Diabetes." *European Journal of Endocrinology* 169 (2013): 725–733.

Nguyen, C. P., M. S. Hirsch, D. Moeny, S. Kaul, M. Mohamoud, and H. V. Joffe. "Testosterone and 'Age-Related Hypogonadism'—FDA Concerns." *New England Journal of Medicine* 378 (2015): 8.

Nieschlag, E., and S. Nieschlag. "Testosterone Deficiency: A Historical Perspective." *Asian Journal of Andrology* 16 (2014): 161–168.

Okun, M. S., H. H. Fernandez, R. L. Rodriguez, J. Romrell, M. Suelter, S. Munson, E. D. Louis, et al. "Testosterone Therapy in Men with Parkinson Disease Results of the TEST-PD Study." *Archives of Neurology* 63 (2006): 729–735.

Okum, M. S., W. M. McDonald, and M. R. DeLong. "Refractory Nonmotor Symptoms in Male Patients with Parkinson Disease Due to Testosterone Deficiency: A Common Unrecognized Comorbidity." *Archives of Neurology* 59, no. 5 (2002): 807–811.

Okun, M. S., B. L. Walter, W. M. McDonald, J. L. Tenover, J. Green, J. L. Juncos, and M. R. DeLong. "Beneficial Effects of Testosterone Replacement for the Nonmotor Symptoms of Parkinson Disease." *Archives of Neurology* 59, no. 11 (2002): 1750–1753.

Page, S. T., L. Hirano, J. Gilchriest, M. Dighe, J. K. Amory, B. T. Marck, and A. M. Matsumoto. "Dutasteride Reduces Prostate Size and Prostate Specific Antigen in Older Hypogonadal Men with Benign Prostatic Hyperplasia Undergoing Testosterone Replacement Therapy." *Journal of Urology* 186, no. 1 (2011): 191–197. doi:10.1016/j.juro.2011.03.026.

Pastuszak, A. W., L. P. Gomez, J. M. Scovell, M. Khera, D. J. Lamb, and L. I. Lipshultz. "Comparison of the Effects of Testosterone Gels, Injections, and Pellets on Serum Hormones, Erythrocytosis, Lipids, and Prostate-Specific Antigen." *Sexual Medicine* 3 (2015): 165–173.

Pastuszak, A. W., H. Mittakanti, J. S. Liu, L. Gomez, L. I. Lipshultz, and M. Khera. "Pharmacokinetic Evaluation and Dosing of Subcutaneous Testosterone Pellets." *Journal of Andrology* 33, no. 5 (2012): 927–937.

Patrick, S. J., and C. C. Carson. "Testosterone Replacement Therapy in Men with Prostate Cancer: What Is the Evidence?" *Sexual Medicine Review* 1, no. 3 (2013): 135–142. doi:10.1002/smrj.15.

Permpongkosol, S., K. Khupulsup, S. Leelaphiwat, S. Pavavattananusorn, S. Thongpradit, and T. Petchthong. "Effects of 8-Year Treatment of Long-Acting Testosterone Undecanoate on Metabolic Parameters, Urinary Symptoms, Bone Mineral Density, and Sexual Function in Men with Late-Onset Hypogonadism." *Journal of Sexual Medicine* 13 (2016): 1199–1211.

Pierorazio, P. M., L. Ferrucci, A. Kettermann, D. L. Longo, E. J. Metter, and H. B. Carter. "Serum Testosterone Is Associated with Aggressive Prostate Cancer in Older Men: Results from the Baltimore Longitudinal Study of Aging." *BJU International* 105, no. 6 (2010): 824–829. doi:10.1111/j.1464-410X.2009.08853.x.

Pinsky, M. R., and W. J. G. Hellstrom. "Hypogonadism, ADAM, and Hormone Replacement." *Therapeutic Advances in Urology* 2, no. 3 (2010): 99–104. doi:10.1177/1756287210369805.

Pintana, H., W. Pongkan, W. Pratchayasakul, N. Chattipakorn, and S. C. Chattipakorn. "Testosterone Replacement Attenuates Cognitive Decline in Testosterone-Deprived Lean Rats, but Not in Obese Rats, by Mitigating Brain Oxidative Stress." *Age* 37 (2015): 84. doi:10.1007/s11357-015-9827-4.

Piszczek, J., M. Mamdani, T. Antoniou, D. N. Juurlink, and T. Gomes. "The Impact of Drug Reimbursement Policy on Rates of Testosterone Replacement Therapy among Older Men." *PLoS ONE* 9, no. 7 (2014): e98003. doi:10.1371 /journal.pone.0098003.

Rabijewski, M., L. Papierska, R. Kuczerowski, and P. Piątkiewicz. "Hormonal Determinants of the Severity of Andropausal and Depressive Symptoms in Middle-Aged and Elderly Men with Prediabetes." *Clinical Interventions in Aging* 10 (2015): 1381–1391.

Rabijewski, M., L. Papierska, and P. Piątkiewicz. "The Relationships between Anabolic Hormones and Body Composition in Middle-Aged and Elderly Men with Prediabetes: A Cross-Sectional Study." *Journal of Diabetes Research* 2016 (2016): 1747261. doi:10.1155/2016/1747261.

Ramasamy, R., J. M. Armstrong, and L. I. Lipshultz. "Preserving Fertility in the Hypogonadal Patient: An Update." *Asian Journal of Andrology* 17 (2015): 197–200.

Ramasamy, R., J. Scovell, M. Mederos, R. Ren, L. Jain, and L. Lipshultz. "Association between Testosterone Supplementation Therapy and Thrombotic Events in Elderly Men." *Urology* 86, no. 2 (2015): 283–286. doi:10.1016 /j.urology.2015.03.049.

Rao, P. K., S. L. Boulet, A. Mehta, J. Hotaling, M. L. Eisenberg, S. C. Honig, L. Warner, D. M. Kissin, A. K. Nangia, and L. S. Ross. "Trends in Testosterone Replacement Therapy Use from 2003 to 2013 among Reproductive-Age Men in the United States." *Journal of Urology* 197, no. 4 (2017): 1121–1126.

Rasmussen, J. J., C. Selmer, P. B. Østergren, K. B. Pedersen, M. Schou, F. Gustafsson, J. Faber, A. Juul, and C. Kistorp. "Former Abusers of Anabolic Androgenic Steroids Exhibit Decreased Testosterone Levels and Hypogonadal Symptoms Years after Cessation: A Case-Control Study." *PLoS ONE* 11, no. 8: e0161208. doi:10.1371/journal.pone.0161208.

Raynaud, J.-P., J. Gardette, J. Rollet, and J.-J. Legros. "Prostate-Specific Antigen (PSA) Concentrations in Hypogonadal Men during 6 Years of Transdermal Testosterone Treatment." *BJU International* 111 (2013): 880–890. doi:10.1111/j.1464-410X.2012.11514.x.

Ready, R. E., J. Friedman, J. Grace, and H. Fernandez. "Testosterone Deficiency and Apathy in Parkinson's Disease: A Pilot Study." *Journal of Neurology and Neurosurgical Psychiatry* 75 (2004): 1323–1326. doi:10.1136/jnnp.2003.032284.

Rodgers, S., M. G. Holtforth, M. P. Hengartner, M. Müller, A. A. Aleksandrowicz, W. Rössler, and V. Ajdacic-Gross. "Serum Testosterone Levels and Symptom-Based Depression Subtypes in Men." *Frontiers in Psychiatry* 6 (2015): 61. doi:10.3389/fpsyt.2015.00061.

Ross, R. J. M., A. Jabbar, T. H. Jones, B. Roberts, K. Dunkley, J. Hall, A. Long, H. Levine, and D. R. Cullen. "Pharmacokinetics and Tolerability of a Bioadhesive Buccal Testosterone Tablet in Hypogonadal Men." *European Journal of Endocrinology* 150 (2004): 57–63.

Rubinow, K. B., C. N. Snyder, J. K. Amory, A. N. Hoofnagle, and S. T. Page. "Acute Testosterone Deprivation Reduces Insulin Sensitivity in Men." *Clinical Endocrinology* 76, no. 2 (2012): 281–288. doi:10.1111/j.1365-2265.2011.04189.x.

Rubinow, K. B., T. Vaisar, C. Tang, A. M. Matsumoto, J. W. Heinecke, and S. T. Page. "Testosterone Replacement in Hypogonadal Men Alters the HDL Proteome but Not HDL Cholesterol Efflux Capacity." *Journal of Lipid Research* 53 (2012): 1376–1383.

Saad, F., A. Aversa, A. M. Isidori, L. Zafalon, M. Zitzmann., and L. Gooren. "Onset of Effects of Testosterone Treatment and Time Span until Maximum Effects Are Achieved." *European Journal of Endocrinology* 165 (2011): 675–685.

Saad, F., A. Haider, G. Doros, and A. Traish. "Long-Term Treatment of Hypogonadal Men with Testosterone Produces Substantial and Sustained Weight Loss." *Obesity* 21 (2013): 1975–1981. doi:10.1002/oby.20407.

Saad, F., A. Yassin, Y. Almehmadi, G. Doros, and L. Gooren. "Effects of Long-Term Testosterone Replacement Therapy, with a Temporary Intermission, on Glycemic Control of Nine Hypogonadal Men with Type 1 Diabetes Mellitus: A Series of Case Reports." *Aging Male* 18, no. 3 (2015): 164–168. doi:10.3109/13685538.2015.1034687.

Saad, F., A. Yassin, G. Doros, and A. Haider. "Effects of Long-Term Treatment with Testosterone on Weight and Waist Size in 411 Hypogonadal Men with Obesity Classes I–III: Observational Data from Two Registry Studies." *International Journal of Obesity* 40 (2016): 162–170.

Saad, F., A. Yassin, A. Haider, G. Doros, and L. Gooren. "Elderly Men over 65 Years of Age with Late-Onset Hypogonadism Benefit as Much from Testosterone Treatment as Do Younger Men." *Korean Journal of Urology* 56 (2015): 310–317. doi:10.4111/kju.2015.56.4.310.

Salam, R., A. S. Kshetrimayum, and R. Keisam. "Testosterone and Metabolic Syndrome: The Link." *Indian Journal of Endocrinology and Metabolism* 16, suppl. 1 (2012): S12–S19. doi:10.4103/2230-8210.94248.

Sarosdy, M. F. "Testosterone Replacement for Hypogonadism after Treatment of Early Prostate Cancer with Brachytherapy." *Cancer* 109, no. 3 (2007): 536–514. doi:10.1002/cncr.22438.

Schnabel, P. G., W. Bagchus, H. Lass, T. Thomsen, and T. B. P. Geurts. "The Effect of Food Composition on Serum Testosterone Levels after Oral Administration of Andriol Testocaps." *Clinical Endocrinology* 66 (2007): 579–585.

Schröder, F. H., J. Hugosson, M. J. Roobol, T. L. J. Tammela, S. Ciatto, V. Nelen, M. Kwiatkowski, et al., for the ERSPC Investigators. "Prostate-Cancer Mortality at 11 Years of Follow-Up." *New England Journal of Medicine* 366 (2012): 981–990.

Schwarz, E. R., and R. D. Willix Jr. "Impact of a Physician-Supervised Exercise-Nutrition Program with Testosterone Substitution in Partial Androgen-Deficient Middle-Aged Obese Men." *Journal of Geriatric Cardiology* 8 (2011): 201–206.

Seara de Andrade, E., Jr., R. Clapauch, and S. Buksman. "Short Term Testosterone Replacement Therapy Improves Libido and Body Composition." *Arquivos Brasileiros de Endocrinologia y Metabologia* 53 (2009): 8.

Sharma, R., O. A. Oni, G. Chen, M. Sharma, B. Dawn, R. Sharma, D. Parashara, V. J. Savin, R. S. Barua, and K. Gupta. "Association between Testosterone Replacement Therapy and the Incidence of DVT and Pulmonary Embolism: A Retrospective Cohort Study of the Veterans Administration Database." *Chest* 150, no. 3 (2016): 563–571.

Shores, M. M., N. L. Smith, C. W. Forsberg, B. D. Anawalt, and A. M. Matsumoto. "Testosterone Treatment and Mortality in Men with Low Testosterone Levels." *Journal of Clinical Endocrinology and Metabolism* 97, no. 6 (2012): 2050–2058.

Shoskes, D. A., B. Tucky, and A. S. Polackwich. "Improvement of Endothelial Function following Initiation of Testosterone Replacement Therapy." *Translational Andrology Urology* 5, no. 6 (2016): 819–823. doi:10.21037 /tau.2016.08.04.

Shoskes, J. J., M. K. Wilson, and M. L. Spinner. "Pharmacology of Testosterone Replacement Therapy Preparations." *Translational Andrology and Urology* 5, no. 6 (2016): 834–843. doi:10.21037/tau.2016.07.10.

Sharma, R., O. A. Oni, K. Gupta, G. Chen, M. Sharma, B. Dawn, R. Sharma, et al. "Normalization of Testosterone Level Is Associated with Reduced Incidence of Myocardial Infarction and Mortality in Men." *European Heart Journal* 36 (2015): 2706–2715. doi:10.1093/eurheartj/ehv346.

Sharma, R., O. A. Oni, K. Gupta, M. Sharma, R. Sharma, V. Singh, D. Parashara, et al. "Normalization of Testosterone Levels after Testosterone Replacement Therapy Is Associated with Decreased Incidence of Atrial Fibrillation." *Journal of the American Heart Association* 6, no. 5 (2017): e004880. doi:10.1161 /JAHA.116.004880.

Shoskes, D. A., Y. Barazani, K. Fareed, and E. Sabanegh Jr. "Outcomes of Prostate Biopsy in Men with Hypogonadism Prior or During Testosterone Replacement Therapy." *International Brazilian Journal of Urology* 41 (2015): 1167–1171. doi:10.1590/S1677-5538.IBJU.2014.0528.

Sicotte, N. L., B. S. Giesser, V. Tandon, R. Klutch, B. Steiner, A. E. Drain, D. W. Shattuck, et al. "Testosterone Treatment in Multiple Sclerosis: A Pilot Study." *Archives of Neurology* 64 (2007): 683–688.

Sinclair, M., M. Grossmann, P. J. Gow, and P. W. Angus. "Testosterone in Men with Advanced Liver Disease: Abnormalities and Implications." *Journal of Gastroenterology and Hepatology* 30 (2015): 244–251. doi:10.1111/jgh.12695.

Singer, N. "Selling That New Man Feeling." *New York Times*, Nov. 23, 2013.

Snyder, G., and D. A. Shoskes. "Hypogonadism and Testosterone Replacement Therapy in End-Stage Renal Disease (ESRD) and Transplant Patients." *Translational Andrology and Urology* 5, no. 6 (2016): 885–889. doi:10.21037 /tau.2016.08.01.

Snyder, P. J., S. Bhasin, G. R. Cunningham, A. M. Matsumoto, A. J. Stephens-Shields, J. A. Cauley, T. M. Gill, et al., for the Testosterone Trials Investigators. "Effects of Testosterone Treatment in Older Men." *New England Journal of Medicine* 374, no. 7 (2016): 611–664. doi:10.1056/NEJMoa1506119.

Takao, T., A. Tsujimura, J. Nakayama, Y. Matsuoka, Y. Miyagawa, S. Takada, N. Nonomura, and A. Okuyama. "Lower Urinary Tract Symptoms after Hormone Replacement Therapy in Japanese Patients with Late-Onset Hypogonadism: A Preliminary Report." *International Journal of Urology* 16 (2009): 212–214. doi:10.1111/j.1442-2042.2008.02202.x.

Taylor, S. R., L. M. Meadowcraft, and B. Williamson. "Prevalence, Pathophysiology, and Management of Androgen Deficiency in Men with Metabolic Syndrome, Type 2 Diabetes Mellitus, or Both." *Pharmacotherapy* 35, no. 8 (2015): 780–792. doi:10.1002/phar.1623.

Traish, A. "Testosterone Therapy in Men with Testosterone Deficiency: Are We beyond the Point of No Return?" *Investigative and Clinical Urology* 57 (2016): 384–400. doi:10.4111/icu.2016.57.6.384.

Traish, A. M., A. Haider, G. Doros, and F. Saad. "Long-Term Testosterone Therapy in Hypogonadal Men Ameliorates Elements of the Metabolic Syndrome: An Observational, Long-Term Registry Study." *International Journal of Clinical Practice* 68, no. 3 (2014): 314–329. doi:10.1111/ijcp.12319.

Traish, A. M., A. Haider, K. S. Haider, G. Doros, and F. Saad. "Long-Term Testosterone Therapy Improves Cardiometabolic Function and Reduces Risk of Cardiovascular Disease in Men with Hypogonadism: A Real-Life Observational Registry Study Setting Comparing Treated and Untreated (Control) Groups." *Journal of Cardiovascular Pharmacology and Therapeutics* 22, no. 5 (2017): 1–20. doi:10.1177/1074248417691136.

Trinick, T. R., M. R. Feneley, H. Welford, and M. Carruthers. "International Web Survey Shows High Prevalence of Symptomatic Testosterone Deficiency in Men." *Aging Male* 14, no. 1 (2011): 10–15.

Uchoa, M. F., V. A. Moser, and C. J. Pike. "Interactions between Inflammation, Sex Steroids, and Alzheimer's Disease Risk Factors." *Frontiers in Neuroendocrinology* 43 (2016): 60–92. doi:10.1016/j.yfrne.2016.09.001.

Ullah, M. I., D. M. Riche, and C. A. Koch. "Transdermal Testosterone Replacement Therapy in Men." *Drug Design, Development and Therapy* 8 (2014): 101–112.

U.S. Food and Drug Administration. "FDA Drug Safety Communication: FDA Evaluating Risk of Stroke, Heart Attack and Death with FDA-Approved Testosterone Products." Jan. 31, 2014. fda.gov/drugs/drugsafety/ucm383904.htm.

U.S. Food and Drug Administration, Department of Health and Human Services. "Androgel."

U.S. Food and Drug Administration, Bone, Reproductive and Urologic Drugs Advisory Committee (BRUDAC). "Advisory Committee Industry Briefing Document: Testosterone Replacement Therapy. Committee Meeting, Sept. 17, 2014.

Vaidya, D., S. H. Golden, N. Haq, S. R. Heckbert, K. Liu, and P. Ouyang. "Association of Sex Hormones with Carotid Artery Distensibility in Men and Postmenopausal Women Multi-Ethnic Study of Atherosclerosis." *Hypertension* 65 (2015): 1020–1025. doi:10.1161/HYPERTENSIONAHA.114.04826.

Vaughan, C., F. C. Goldstein, and J. L. Tenover. "Exogenous Testosterone Alone or With Finasteride Does Not Improve Measurements of Cognition in Healthy Older Men with Low Serum Testosterone." *Journal of Andrology* 28, no. 6 (2007): 875–882.

Vigen, R., C. I. O'Donnell, A. E. Baron, G. K. Grunwald, T. M. Maddox, S. M. Bradley, A. Barqawi, et al. "Association of Testosterone Therapy with Mortality, Myocardial Infarction, and Stroke in Men with Low Testosterone Levels." *Journal of the American Medical Association* 310, no. 17 (2013): 1829–1836. doi:10.1001/jama.2013.280386.

Wallis, C. J. D., H. Brotherhood, and P. J. Pommerville. "Testosterone Deficiency Syndrome and Cardiovascular Health: An Assessment of Beliefs, Knowledge and Practice Patterns of General Practitioners and Cardiologists in Victoria, BC." *Canadian Urology Association Journal* 8, nos. 1–2 (2014): 30–33. doi:10.5489/cuaj.1448.

Wallis, C. J. D., K. Lo, Y. Lee, Y. Krakowsky, A. Garbens, R. Satkunasivam, S. Herschorn, et al. "Survival and Cardiovascular Events in Men Treated with Testosterone Replacement Therapy: An Intention-to-Treat Observational Cohort Study." *Lancet Diabetes and Endocrinology* 4 (2016): 498–506. doi:10.1016/S2213-8587(16)00112-1.

Wang, W., T. Jiang, C. Li, J. Chen, K. Cao, L.-W. Qi, P. Li, W. Zhu, B. Zhu, and Y. Chen. "Will Testosterone Replacement Therapy Become a New Treatment of Chronic Heart Failure? A Review Based on 8 Clinical Trials." *Journal of Thoracic Diseases* 8, no. 5 (2016): E269–E277.

Wicks, P. "Hypothesis: Higher Prenatal Testosterone Predisposes ALS Patients to Improved Athletic Performance and Manual Professions." *Amyotrophic Lateral Sclerosis* 13 (2012): 251–253.

Yanovski, S. Z., and J. A. Yanovski. "Long-Term Drug Treatment for Obesity: A Systematic and Clinical Review." *Journal of the American Medical Association* 311, no. 1 (2014): 74–86. doi:10.1001/jama.2013.281361.

Yaron, M., Y. Greenman, J. B. Rosenfeld, E. Izkhakov, R. Limor, E. Osher, G. Shenkerman, K. Tordjman, and N. Stern. "Effect of Testosterone Replacement Therapy on Arterial Stiffness in Older Hypogonadal Men." *European Journal of Endocrinology* 160 (2009): 839–846.

Yassin, A., Y. Almehmadi, F. Saad, G. Doros, and L. Gooren. "Effects of Intermission and Resumption of Long-Term Testosterone Replacement Therapy on Body Weight and Metabolic Parameters in Hypogonadal Middle-Aged and Elderly Men." *Clinical Endocrinology* 84 (2016): 107–114.

Yassin, A., and G. Doros. "Testosterone Therapy in Hypogonadal Men Results in Sustained and Clinically Meaningful Weight Loss." *Clinical Obesity* 3 (2013): 73–83. doi:10.1111/cob.12022.

Yassin, A., J. E. Nettleship, R. A. Talib, Y. Almehmadi, and G. Doros. "Effects of Testosterone Replacement Therapy Withdrawal and Retreatment in Hypogonadal Elderly Men upon Obesity, Voiding Function and Prostate Safety Parameters." *Aging Male* 19, no. 1 (2016): 64–69. doi:10.3109/13685538.2015.1126573.

Yassin, A., M. Salman, R. A. Talib, and D.-J. Yassin. "Is There a Protective Role of Testosterone against High-Grade Prostate Cancer? Incidence and Severity of Prostate Cancer in 553 Patients Who Underwent Prostate Biopsy: A Prospective Data Register." *Aging Male* 20, no. 2 (2017): 125–133. doi:10.1080/13685538.2017.1298584.

Yassin, A. A., J. E. Nettleship, Y. Almehmadi, D.-J. Yassin, Y. El Douaihy, and F. Saad. "Is There a Relationship between the Severity of Erectile Dysfunction and the Comorbidity Profile in Men with Late Onset Hypogonadism?" *Arab Journal of Urology* 13 (2015): 162–168.

Yassin, D.-J., G. Doros, P. G. Hammerer, and A. A. Yassin. "Long-Term Testosterone Treatment in Elderly Men with Hypogonadism and Erectile Dysfunction Reduces Obesity Parameters and Improves Metabolic Syndrome and Health-Related Quality of Life." *International Society for Sexual Medicine* 11 (2014): 1567–1576.

Yassin, D.-J., Y. El Douaihy, A. A. Yassin, J. Kashanian, R. Shabsigh, and P. G. Hammerer. "Lower Urinary Tract Symptoms Improve with Testosterone Replacement Therapy in Men with Late-Onset Hypogonadism: 5-Year Prospective, Observational and Longitudinal Registry Study." *World Journal of Urology* 32 (2014): 1049–1054. doi:10.1007/s00345-013-1187-z.

Zając, A., M. Wilk, T. Socha, A. Maszczyk, and J. Chycki. "Effects of Growth Hormone and Testosterone Therapy on Aerobic and Anaerobic Fitness, Body Composition and Lipoprotein Profile in Middle-Aged Men." *Annals of Agricultural and Environmental Medicine* 21, no. 1 (2014): 156–160.

Zhang, X. W., Z. H. Liu, X. W. Hu, Y. Q. Yuan, W. J. Bai, X. F. Wang, H. Shen, and Y. P. Zhao. "Androgen Replacement Therapy Improves Psychological Distress and Health-Related Quality of Life in Late Onset Hypogonadism Patients in Chinese Population." *Chinese Medical Journal* 125, no. 21 (2012): 3806–3810.

Zhao, J. V., T. H. Lam, C. Jiang, S. S. Cherny, B. Liu, K. K. Cheng, W. Zhang, G. M. Leung, and C. M. Schooling. "A Mendelian Randomization Study of Testosterone and Cognition in Men." *Nature Scientific Reports* 6 (2016): 21306. doi:10.1038/srep21306.

Ziehn, M. O., A. A. Avedisian, S. M. Dervin, E. A. Umeda, T. J. O'Dell, and R. R. Voskuhl. "Therapeutic Testosterone Administration Preserves Excitatory Synaptic Transmission in the Hippocampus during Autoimmune Demyelinating Disease." *Neuroscience* 32, no. 36 (2012): 12312–12324. doi:10.1523/JNEUROSCI.2796-12.2012.

Zonderman, A. B. "Predicting Alzheimer's disease in the Baltimore Longitudinal Study of Aging." *Journal of Geriatric Psychiatry and Neurology* 18, no. 4 (2005): 192–195.